AF577238

PUMPS:
SELECTION, SYSTEMS AND APPLICATIONS

2ND EDITION

GULF PUBLISHING COMPANY
Book Division
Houston, London, Paris, Tokyo

PUMPS:
SELECTION, SYSTEMS AND APPLICATIONS
2ND EDITION

This edition published 1984
by Gulf Publishing Company
Houston, Texas.

Simultaneously published in 1984
by The Trade & Technical Press Limited,
Crown House, Morden, Surrey, SM4 5EW, England.

PRINTED IN THE UNITED KINGDOM

ISBN 0–87201–736–2

Library of Congress Catalog Card No. 83–82095

CONTENTS

1. Types of Fluids

THE BASIC FUNCTION of any pump is to transport a fluid, whence it follows that the performance achieved and the means by which the fluid is acted upon must be closely related to the characteristics of the fluid involved, as well as to the fluid characteristics in general. Rather loosely, a fluid may be defined as a substance with a non-rigid assembly of particles which can be made to flow in bulk by the application of a suitable force, although a distinction must be made between a true liquid and a plastic solid which is capable of cold flow (plastic flow) under stressed conditions. On the other hand, rigid solids may well be capable of flowing like a fluid when they are in the form of a non-rigid assembly, *eg* in powdered or particle form, as well as being capable of cold flow under extreme pressures, when in homogeneous solid form. In the former case a solid may be 'pumpable', in the latter case it may be 'deformable' or 'exturdable' with strictly limited flow possible.

Many parameters concerned with pump performance, pump type selection and material selection are related to the specific properties of individual fluids rather than basic fluid parameters themselves. The following is a summary of the main fluid types.

Water and Aqueous Solutions

Not only is water the most common liquid pumped in terms of volume, but water is virtually a 'standard' fluid for determining pump performance. In other words, particularly in the case of general service pumps, performance figures normally relate to handling clean cold water and may need adjustment to estimate handling products other than water and aqueous solutions, (an exception is gear pumps which are commonly rated on their oil-handling performance).

The effect of dissolved solids (*ie* aqueous solutions) is generally to increase the specific gravity of the fluid slightly and at the same time raise its boiling point (*ie* lower its vapour pressure) and lower its freezing point. The effect on viscosity is generally negligible except where the admixture results in definite thickening. However, dissolved solids may appreciably increase the chemical activity of the product, an overall guide here being the pH value of the aqueous solution.

The low viscosity of water and aqueous solutions can place a premium on seal selection, particularly in higher pressure systems.

Oils

Oil fluids range from very high to very low viscosities, with a particular problem that actual working viscosity can vary widely with temperature. Some oil fuels may also be classed as volatile and/or hazardous fluids.

Particular advantages offered by oil fluids is that most have good lubricating properties and are relatively easy to seal.

Dissolved Liquids

The admixture of two liquids which do not react chemically, normally results in a modification of both the specific gravity and viscosity pro rata to the proportions of the two liquids. The addition of quite small proportions of aggressive substances (such as acids) to water can, however, drastically increase its chemical and electrochemical activity.

Liquid Mixtures

Certain liquids will not mix in the sense of solution, but one may be dispersed through and contained in the other in the form of suspended globules or minute droplets. Such liquid mixtures are generally known as emulsions and may be stable (in which case they will remain permanently mixed), or unstable (in which case they will tend to separate out). The two basic types of emulsion are oil-in-water and water-in-oil, depending on which constituent is the external phase.

Emulsions are generally distinguished by lack of a true viscosity, *ie* a viscosity figure which is constant at any particular temperature. The apparent viscosity, which is generally appreciably higher than that of the external phase component alone, degenerates with shear, such as cause by flow, and if the shear rate is high enough a complete breakdown of the emulsion may occur – see **Fig 1.1**. The viscosity in this case is no greater than that of the external phase.

Fluids with apparent, rather than true viscosities, are generally classified as non-Newtonian.

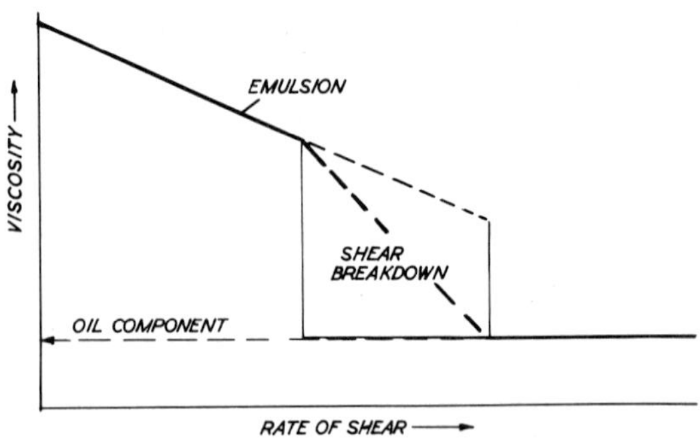

Fig 1.1

Non-Newtonian Fluids

In the case of a conventional or Newtonian fluid the rate of shear is linearly proportional to the rate of stress – viscosity is thus constant provided the pressure and temperature of the fluid remain constant. Fluids which deviate from this basic relationship, whether simple fluids or liquid-solid mixtures, are known as non-Newtonian fluids. These may be further subdivided according to type.

Bingham plastics are non-Newtonian fluids characterised by having a yield point, and flow cannot take place until this yield point is exceeded – see **Fig 1.2**. Such a fluid does not have a definite viscosity, but an *apparent viscosity,* the value of which is not constant but dependent on shear rate.

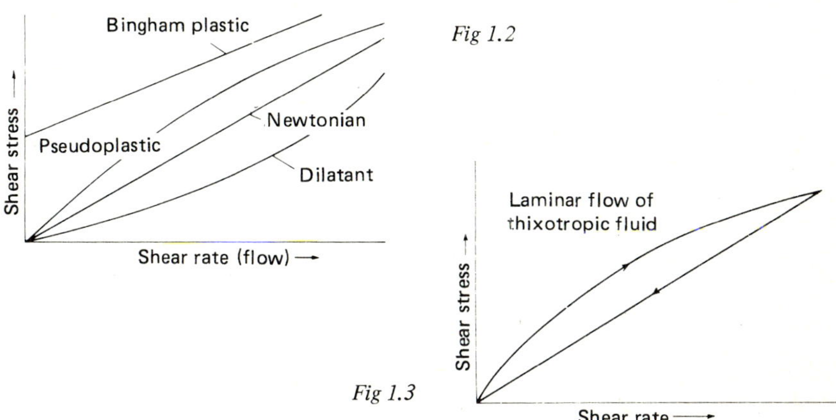

Fig 1.2

Fig 1.3

Intermediately between a true Newtonian fluid and a Bingham plastic are the so-called *pseudo-plastics.* Whilst these have no 'yield point' (and thus differ from a Bingham plastic), the shear stress relationship is non-linear (and thus different from a Newtonian fluid). Again, therefore, such fluids have an *apparent viscosity* which is variable with shear rate.

Quite a number of liquid-solid mixtures exhibit the nature of a Bingham plastic or pseudo-plastic under laminar flow conditions.

A contrasting class of fluid is the so-called *dilatants,* which have the characteristic of decreasing the rate of shear as stress is increased. Their apparent viscosity thus increases, non-linearly, with shear rate. This is because agitation results in a rearrangement of the solids, with a resultant increase in volume. A typical example of a dilatant fluid is a suspension of starch in water.

Further distinct characteristics are exhibited by other types of non-Newtonian fluids. Thus thixotropic fluids exhibit a hysteresis effect in that their apparent or instantaneous viscosity is dependent on the previous history of the fluid. **Fig 1.3** is a typical rheogram for a thixotropic fluid under laminar flow.

Turbulent flow will tend to change the structure of the fluid, which will recover if left standing for a sufficient time.

Chemically Active Fluids

Chemically active fluids can present problems with pump material selection, and also gland design or even pump type selection if it is necessary to eliminate leakage entirely (*eg* in the latter case a double mechanical seal, or a glandless pump may be necessary).

The acidity or alkalinity of such a fluid is expressed by its pH value, or more specifically by actual compatibility with individual materials or *ISO-corrosion charts.*

Slurries and Pastes

Non abrasive slurries and pastes may degrade the water performance of standard pumps, or they may require modified or special designs of pumps to handle these effectively – see **Chapters 6** and **10**.

Liquid/Solid Mixtures

Homogeneous mixtures of solids and liquids may range in character from thin pastes to thick pulps and stock, with flow characteristics varying accordingly. In general, with such mixtures, gravitational flow characteristics remain similar to those of normal fluids up to a particular consistency. With higher consistencies gravitational flow separates into droplets, the size of the droplets increasing with increasing consistency until they take the form of large slugs. These flow characteristics can affect the design of pumps suitable for handling particular mixtures, especially with regard to flow on the intake side.

Solids in Suspension

Fluids containing solids in suspension, also generally known as *slurries,* exhibit an increase in specific gravity and also an increase in viscosity. Such fluids are, however, non-Newtonian and the viscosity figure under any conditions is an apparent rather than a true one. The degree of suspensions, and thus the consistency of the fluid, may also depend on the flow rate. This may be generally related to a *fall velocity* or the minimum flow rate necessary to maintain the solids in suspension and prevent them from settling out. This, in turn, depends on the size of the solids, and also their concentration. Approximate minimum flow velocities to retain solids in suspension for various classes of slurries are:

Fines (particle size 75 μm or less) – 3 ft/sec (6.9 m/sec)
Sands (particle size 75–850 μm) – 5 ft/sec (1.5 m/sec)
Coarse (particle size 850–5 000 μm) – 7 ft/sec (2 m/sec)
Sludge (particle size ¼ in (6 mm) diameter and above) – 11 ft/sec (3.4 m/sec)

These empirical figures are based on a solids content of 30–35% by weight and solids of specific gravity 2.5–3.0.

Although sludges are rated as containing larger solids, true sludges also normally contain a high proportion of finer solids or mud.

The handling of fluids containing solids sets particular problems as far as the pump design is concerned. Thus the pump must have internal clearances sufficient to pass the solids without jamming or clogging, and usually also require hardened surfaces of hard facings to resist the abrading action of such solids.

Sewage

Raw sewage may contain stringy and fibrous solids as well as debris such as pieces of timber, tin cans, old shoes and boots, plastic containers, *etc* – *ie* very much larger solids than contained in slurry. This calls for further modification of pump design/installation to prevent clogging, also larger pipes to pass the solids. An alternative, or additional, solution in such cases may be the fitting of a comminutating device or macerator preceding the pump to reduce the size of the fibrous material before entering the pump.

Delicate Solids

A further class of suspended solids are those where the solid content is delicate and may be damaged by normal pumping action. Where this has to be avoided – *ie* the solids content is an important part of the product being handled – the pump type may have to be selected accordingly to eliminate damage to the product.

Volatile Fluids

The classification *volatile fluids* relates to all liquids which may present special problems in handling because of their volatile nature. This can affect particularly the performance of a centrifugal pump.

Fluids with Dissolved Gases

Most liquids are capable of dissolving a proportion of air or other gases, the actual proportion absorbed by a particular fluid being dependent on temperature and pressure. For example, at normal ambient temperatures water can absorb about 3% of its own volume of air; and light oils up to 8% of their own volume of air.

The presence of absorbed air or gases does not normally affect the characteristics of the fluid except for the fact that absorbed gases will tend to be released if the pressure is decreased, or the temperature is increased, with the possibility of air pockets forming in certain parts of the system. Such troubles can normally be eliminated by avoiding potential air traps in the piping system and/or suitable venting.

Entrained Gases

Gases, particularly air, may also be contained in fluids in the form of bubbles dispersed through the liquid, the mixture of liquid and air being a mechanical one. This can seriously affect the performance of a pump, although certain types are more susceptible to air entrainment than others. Thus rotary pumps will suffer a loss of capacity somewhat greater than the percentage of gas entrainment. The presence of entrained air may also promote foaming, particularly with oil fluids.

Centrifugal pumps in particular do not handle gases readily, since the centrifugal action tends to keep the lighter media at the centre – *ie* impeller eye – and if sufficient gases collect, the flow ceases due to the eye being completely full of gas vapour; the impeller cannot develop sufficient head on the gas, due to its low specific gravity, to force it outwards to create the necessary pressure drop in the impeller eye to achieve liquid flow. Many designs of centrifugal pumps are available which use an internal recirculating method to overcome the above problem and, whilst inefficient, these methods are quite successful when handling small quantities of gas, providing the viscosity is 3 centipoise or less, and the discharge pressure is low. Separate 'vacuum' pumps are often fitted to centrifugal pumps and these are very successful but do create an extra system problem, especially with obnoxious liquids.

Viscous Fluids

Water has viscosity of 1 centistoke. Fluids with higher viscosities up to 100 centistokes can be handled by most types of pumps with little loss of performance. Fluids with a viscosity greater than 100 centistokes may be classified as *viscous fluids* and become increasingly difficult to handle efficiently – see **Chapter 10.** Eventually their flow characteristics approach those of a plastic solid only capable of plastic flow under extreme pressures. Individual pump designs may be needed to handle such products – see **Chapter 6.**

Hazardous Fluids

Fluids which are toxic or inflammable are generally classed as hazardous fluids. Also included under such a heading are fluids which are strongly chemically active and could thus present a hazard if they or their vapours can escape through leakage, *etc.* The latter include most acids, alkalis and strongly reactive chemical solutions.

Hazardous fluids may require the use of special pumps – *eg* to eliminate gland leakage, and flameproof motors – *eg* to eliminate explosion risk with flammable fluids. The handling of inflammable fluids or those which create explosive vapours must also conform to the National regulations which may apply, such as:

USA – National Fire Protection Association.

Britain – HM Factory Inspectorate.

Continental Europe – VDE specifications and PTB approved controls.

Switzerland – SEV test certificate.

Table I lists a number of fluids classified as 'hazardous' and the degree of hazard rating for toxicity and flammability.

TABLE I – TOXIC AND FLAMMABLE FLUIDS

Group (BS.229)	Fluid	Spontaneous Ignition Temperature		Hazard Rating*	
		Deg.F	Deg.C	Toxicity	Flammability
I	Methane	999	538	1	4
II	Blastfurnace gas	–	–	2	4
	Carbon monoxide	1 128	610	2	4
	Propane	871	466	1	4
	Butane	761	405	1	4
	Pentane	588	308	1	4
	Hexane	502	261	1	3
	Heptane	433	222	1	3
	Iso-octane	428	220	0	3
	Decane	406	208	0	2
	Benzene	1 044	564	2	3
	Xylene	867	464	2	3
	Cyclohexane	500	260	1	3
	Acetone	1 000	538	1	3
	Ethyl methyl ketone (MEK)	960	516	1	3
	Methyl acetate	935	502	1	3
	Ethyl acetate	800	427	1	3
	n-Propyl acetate	842	450	1	3
	n-Butyl acetate	790	422	1	3
	Amyl acetate	–	–	1	3
	Chloroethylene (vinyl chloride)	882	472	2	4
	Methanol (methyl alcohol)	867	464	1	3
	Ethanol (ethyl alcohol)	793	423	0	3
	i-Butyl alcohol	800	427	1	3
	n-Butanol (butyl alcohol)	689	365	1	3
	Amyl alcohol	572	300	1	3
	Ethyl nitrite	194	90	2	4
	Buta 1 : 3-diene	804	428	2	4
IIIa	Ethylene	842	450	1	4
	Diethyl ether (ethyl ether)	356	180	2	4
	Ethylene oxide	804	428	2	4
IIIb	Coal gas (town gas)	–	–	2	4
	Coke oven gas	–	–	2	4
IV	Acetylene	571	294	1	4
	Carbon disulphide	212	100	2	3
	Ethyl nitrate	–	–	–	3
	Hydrogen	1 085	580	0	4
	Water gas	–	–	2	4

*Degree of hazard from 0 = non-existent to 4 = very hazardous

2. Fluid Parameters

THE MAJORITY of pumping applications are concerned with the handling of true fluids, and here a distinction must be drawn between gases and liquids. Both are fluids, by definition, with the same parameters which affect fluid flow (and thus pumping characteristics) but the values of these parameters are widely different. In particular, gases are compressible, whilst liquids may be generally regarded as incompressible up to quite high pressures (compressibility effects on liquids only becoming apparent at pressures of the order of 1 000 lb/in^2 (70 bar) and above). Thus the transport of gases normally involves a degree of compression (and the pump itself may, indeed, be a true compressor) or a substantial fluid volume change (as in the case of vacuum pumps). Liquid transport is normally achieved without volume change (except that elasticity effects may be introduced into the system at high pressures) and performance is more readily predeterminable. For practical purposes, therefore, the handling of gases and liquids can be considered as separate pumping problems, although certain types of pumps may be suitable for handling both classes of fluids (*eg* see **Table I**).

TABLE I – FLUID HANDLING CAPABILITIES OF PUMPS

Pumps for Liquids Only	Pumps for Liquids or Gases	Pumps for Air and Gases Only
Centrifugal Mixed flow Regenerative Axial flow Submersibles Flexible impeller Eccentric screw Archimedean screw Gear Lobe rotor Diaphragm Peristaltic	Reciprocating Screw Vane Centrifugal* Diaphragm*	Roots Liquid ring

*Not a normal choice for handling air or gases, but centrifugal pumps may be used for high-volume, low pressure compressor duties (blowers); and diaphragm pumps may be used for low-volume duties with gases.

The chief fluid engineering parameters concerned with the handling of liquids are *specific gravity* and *viscosity*. Both affect the head and capacity (or speed) at which the pump can operate, and the power input required. Viscosity in particular (and specific gravity to a lesser extent) is also temperature dependent.

Specific Gravity

Specific gravity is a dimensionless quantity and is defined as the ratio of the specific weight of the liquid to the specific weight of water, *viz:*

$$\text{Specific gravity} = \frac{w_{fluid}}{w_{water}}$$

The weight per unit volume of a fluid is not constant but variable with temperature, and to a much lesser extent with pressure. The latter can generally be ignored in pumping problems. Specific gravity, however, must be defined relative to the weight of water at a specific temperature, *viz* 60°F or 15°C (see also **Table II**). The specific gravity (SG) of any liquid is thus suitably defined as –

$SG = 0.016\, w_L$ (where w_L is the specific weight or density of the liquid in lb/ft^3)

$SG = w_L$ (where w_L is the specific weight or density of the liquid in kg/lit).

Metric practice quotes specific gravity (or relative density) as $SG_{15/4°C}$ where the specific weight of the fluid is determined at 15°C and the standard weight of water at 4°C. In English units specific gravity is quoted as $SG_{60/60°F}$ (*ie* both fluid weight and water weight at 60°F). The two are not equivalent, although they are usually assumed so for most practical purposes. If correction between the two systems of units is required the appropriate formulas are –

$$SG_{60/60°F} = SG_{15/4°C} + \frac{SG_{15/4°C}}{1\,000}$$

$$SG_{15/4°C} = SG_{60/60°F} - \frac{SG_{60/60°F}}{1\,000}$$

Note: these formulas are approximate rather than exact.

Other Hydrometer Scales

Other empirical hydrometer (relative density) scales in use include API degrees for oil fluids; Baumé degrees (US) and Twaddell degrees (UK) – see **Table III** for conversions; also **Table IV** for other empirical scales.

Note – in Table II

* Specific weight or density

‡ 62.366 weighed in vacuo

Note: weighings for 30°C upwards are nominal

† 62.426 weighed in vacuo

62.316 weighed in vacuo

ϕ 1.00 weighed in vacuo

TABLE II – WEIGHT OF CLEAN WATER

Temperature °F	Temperature °C	Weight in Pounds per cu ft*	Weight in Pounds per Imperial Gallon	Weight in Pounds per US Gallon	Weight in Kilograms per Litre	SG (nominal)
–	4	62.356 †	10.0109	8.3359	0.99888	1.0
–	15	62.305	10.0026	8.3290	0.99805	1.0
60	–	62.299‡	10.0018	8.3283	0.99796	1.0
62	–	62.288	10.0000	8.3268	0.99779	1.0
–	20	62.250#	9.9939	8.3217	0.99717	1.0
–	25	62.179	9.9829	8.3122	0.99603	1.0
–	30	62.15	9.98	8.3	0.99	0.99
–	40	62.0	9.95	8.3	0.99	0.99
–	60	61.4	9.85	8.2	0.98	0.95
–	80	60.6	9.75	8.1	0.97	0.97
212	100	59.8	9.6	8.0	0.96	0.96

TABLE IV – MISCELLANEOUS HYDROMETER SCALES
(N = scale divisions)

Name	Standard Temp	Specific Gravity: Liquids Heavier than Water	Specific Gravity: Liquids Lighter than Water
Barkometer	–	$1\,000 + \frac{N}{1\,000}$	–
Baume (original)	150°C	$\frac{146.3}{146.3 - {}^\circ\text{Baume}}$	$\frac{146.3}{136.3 + {}^\circ\text{Baume}}$
Baume (American)	60°F	$\frac{145}{145 - {}^\circ\text{Baume}}$	$\frac{145}{130 + {}^\circ\text{Baume}}$
Baume (German)	150°C	$\frac{144.3}{144.3 - {}^\circ\text{Baume}}$	$\frac{144.3}{144.3 + {}^\circ\text{Baume}}$
Beck	125°C	$\frac{170}{170 - N}$	$\frac{170}{170 + N}$
Brix	–	–	$\frac{400}{400 - N}$
Gay-Lussac	–	–	$\frac{100}{100 - N}$
Twaddell	–	$1\,000 + \frac{{}^\circ\text{Twaddell}}{200}$	–

TABLE III – EMPIRICAL HYDROMETER SCALE CONVERSIONS

Original	Conversion to	Formula
$SG_{60/60°F}$	$SG_{15/4°C}$	$= SG_{60/60°F} - \frac{SG_{60/60°F}}{1\,000}$
	$SG_{20/20°C}$	$= SG_{60/60°F} + \frac{SG_{60/60°F}}{1\,500}$
	API degrees	$= \frac{141.5}{SG_{60/60°F}} - 131.5$
	Baume degrees (US)	$= \frac{140}{SG_{60/60°F}} - 130$ (liquids lighter than water)
		$= 145 - \frac{145}{SG_{60/60°F}}$ (liquids heavier than water)
$SG_{15/4°F}$	$SG_{60/60°F}$	$= SG_{15/4°C} + \frac{SG_{15/4°C}}{1\,000}$
	$SG_{20/20°C}$	$= SG_{15/4°C} + \frac{SG_{15/4°C}}{600}$
	API degrees	$= \frac{141.36}{SG_{15/4°C}} - 131.5$
	Twaddell degrees	$= \frac{SG_{15/4°C} - 1.0}{0.005}$
$SG_{20/20°C}$	$SG_{60/60°F}$	$= SG_{20/20°C} - \frac{SG_{20/20°C}}{1\,500}$
	$SG_{15/4°C}$	$= SG_{20/20°C} - \frac{SG_{20/20°C}}{600}$
API Degrees	$SG_{60/60°F}$	$= \frac{141.5}{°API + 131.5}$
	$SG_{15/4°C}$	$= \frac{141.36}{°API + 131.5}$

Density

Density is defined as mass per unit where the ISO Standard unit is kg/m^3 and the English/American unit lb/ft^3.

1 kg/m^3 = 0.6243 lb/ft^3 (or 1/16 x lb/ft^3 close approx)

1 lb/ft^3 = 16.018 kg/m^3 (or 16 kg/m^3 close approx)

Metric density may also be quoted in units of metric tonnes/m^3, when

1 tonne/m^3 = 62.428 lb/ft^3

1 lb/ft^3 = 0.016 tonnes/m^3

Note that the same numerical value for (metric) density is given by the units tonnes/m^3, kg/dm^3 and grams/cm^3.

Some confusion can arise from the fact that the French word *densité* means *specific gravity* (not density). The French for density is *mass volumique.*

TABLE V – SPECIFIC GRAVITY/DENSITY CORRECTION COEFFICIENTS

Fluid	Coefficient	
	per °F	per °C
Acetone	0.00062	0.00111
Alcohol: General	0.00044	0.00079
Ethyl	0.00048	0.00087
Methyl	0.00052	0.00094
Benzene	0.00058	0.00105
Benzole	0.00054	0.00098
Coal Tar, crude	0.00036	0.00065
Creosote	0.00042	0.00075
Ether (Isopropyl)	0.00057	0.00102
Methyl Isobutyl Carbonol	0.00045	0.00081
Methyl Isobutyl Ketone	0.00053	0.00095
Methyl Ethyl Ketone	0.00057	0.00103
Naphtha	0.00047	0.00085
Toluol	0.00050	0.00090
Xylol	0.00048	0.00086
Petroleum Products:*		
SG 0.6–0.62	0.00056	0.00101
0.62–0.64	0.00054	0.00098
0.64–0.66	0.00052	0.00094
0.66–0.68	0.00050	0.00090
0.68–0.72	0.00048	0.00085
0.72–0.74	0.00046	0.00081
0.74–0.76	0.00044	0.00078
0.76–0.78	0.00042	0.00076
0.78–0.82	0.00040	0.00072
0.82–0.86	0.00038	0.00067
0.86–0.93	0.00036	0.00065
0.93–1.025	0.00035	0.00063
1.025–1.075	0.00034	0.00061
1.075–1.125	0.00033	0.00059

*Consult ASTM/IP Petroleum Measurement Tables for more accurate data.

Specific Gravity and Temperature

Specific gravity/density variations with temperature may be significant, particularly where wide temperature ranges and large volumes of liquid are involved. In the case of water the variation is normally small enough to be ignored (see **Table IV**). In the case of petroleum products and related fluids, specific gravity/density correction for temperature can be derived from standard formulas –

$SG_T = SG_{60°} - C_F \times SG_{60°} \times (T - 60)$ (Fahrenheit temperature)

$SG_T = SG_{15°} - C_C \times SG_{15°} \times (T - 15)$ (Centigrade temperature)

$W_T = w_{60°} - C_F \times w_{60°} \times (T - 60)$ (Fahrenheit temperature)

$W_T = w_{15°} - C_C \times w_{15°} \times (T - 15)$ (Centigrade temperature)

where:

SG_T = specific gravity at temperature T

W_T = specific weight at temperature T

C_F = correction coefficient per °F – see **Table V**

C_C = correction coefficient per °C - see **Table V**

The suffixes refer to the standard temperatures for specific gravity and specific weight in the respective formulas.

Volume Changes with Temperature

Volume changes with temperature can be estimated in a similar manner, *viz* –

$V_T = V_{60} - v_F \times V_{60} \times (T - 60)$ (Fahrenheit temperature)

$V_T = V_{15} - v_C \times V_{15} \times (T - 60)$ (Centigrade temperature)

where:

v_F = volume correction coefficient per °F – see **Table VI**

v_C = volume correction coefficient per °C – see **Table VI.**

Note, however, that in this case if the specific gravity of an oil fluid is measured at a higher temperature than 60°F (or 15°C) it must first be corrected to find the equivalent specific gravity at either 60°F or 15°C to enter **Table VI** to determine the true volume correction coefficient.

Volume correction coefficients can also be used more simply to determine the percentage ullage required for drum filling – assuming that the drums are filled at normal ambient temperature but may be expected to experience a maximum temperature rise of t degrees during storage or transit. The likely expansion, and thus the percentage ullage required, can be calculated as 100 x V_T x t volume correction coefficient for the fluid concerned.

TABLE VI – VOLUME CORRECTION COEFFICIENTS

Fluid		Coefficient per °F	Coefficient per °C
Acetic Acid		0.0006	0.00107
Acetone		0.00079	0.00143
Alcohol:	Amyl	0.00052	0.00093
	Ethyl	0.00061	0.00110
	Methyl	0.00068	0.00122
Aniline		0.00047	0.00085
Benzene		0.00069	0.00124
Carbon Disulphide		0.00067	0.00121
Chloroform		0.00071	0.00126
Ethyl Ether		0.00091	0.00163
Isopropyl Ether		0.00080	0.00144
Ethyl Bromide		0.00076	0.00137
Glycerine		0.00030	0.00053
Methyl Ethyl Ketone		0.00073	0.00131
Methyl Isobutyl Carbinol		0.00057	0.00103
Methyl Iodide		0.00067	0.00121
Olive Oil		0.00039	0.00070
Paraffin up to 20°C		0.00056	0.00090
	20–199°C	0.00061	0.00110
Pentane		0.00089	0.00159
Sulphuric Acid (100%)		0.00032	0.00057
Toluene		0.00061	0.00109
Turpentine		0.00052	0.00094
Xylol		0.00056	0.00101
Petroleum Products		0.0003–0.0010*	0.0006–0.0017*
Water:	5–10°C	0.00003	0.000053
	10–20°C	0.000084	0.000150
	20–40°C	0.000167	0.000302
	40–60°C	0.000255	0.000458
	60–80°C	0.00033	0.000587

*Depending on Specific Gravity – consult ASTM/IP Petroleum Measurement Tables

Viscosity

Dynamic viscosity (μ) is measured in terms of (force x time)/area, the standard unit being *centipoise* where

$$1 \text{ centipoise} = 1 \text{ Newton sec/m}^2 \times 10^{-3}$$

$$= 0.021 \text{ lb sec/ft}^2 \text{ (approx).}$$

Kinematic viscosity (ν) is measured in terms of area/time and is normally used for engineering calculations. The standard unit is the *centistoke* where

$$1 \text{ centistoke} = \text{m}^2\text{sec} \times 10^{-6}$$

$$= 0.001549 \text{ m}^2\text{/sec}$$

$$= 0.000108 \text{ ft}^2\text{/sec}$$

In practice, fluid viscosity is usually measured by standard viscometers, which have arbitrary scales. The most common method is measurement of the time of flow of a specific volume of fluid through a standard jet, viscosity thus being rendered in terms of seconds. Such scales have no exact equivalent to readings obtained on other standard instruments, nor are they capable of exact conversion to standard units. Conversions close enough for most purposes can, however, be obtained from viscosity conversion tables or by calculation (see **Table VII**).

TABLE VII – APPROXIMATE VISCOSITY CONVERSIONS: ARBITRARY SCALES

Scale	Multiplier	Approx. Equivalent
Redwood No 1	1.14 0.033 0.25	Saybolt Universal Seconds Engler Degrees Centistokes*
Redwood No 2	10	Redwood No 1
Saybolt (SUS)	0.88 0.0285 0.22	Redwood No 1 Seconds Engler Degrees Centistokes**
Engler Degrees (E°)	30.7 35 7.6	Redwood No 1 Seconds Saybolt Universal Seconds Centistokes***

Note: General conversion factors are:

$$\text{*Centistokes} = \left(0.264T - \frac{190}{T}\right) \text{ for } T = 40\text{–}85 \text{ seconds}$$

$$= \left(0.247T - \frac{65}{T}\right) \text{ for } T \text{ over } 85 \text{ seconds}$$

$$\text{**Centistokes} = \left(0.226T - \frac{195}{T}\right) \text{ for } T \text{ less than } 100 \text{ seconds}$$

$$= \left(0.220T - \frac{135}{T}\right) \text{ for } T \text{ over } 100 \text{ seconds}$$

$$\text{***Centistokes} = 7.6 \left(1 - \frac{1}{E^{\circ}}\right) \times E^{\circ}$$

Viscosity/Temperature Effects

Viscosity-temperature characteristics are most conveniently plotted on an ASTM chart, using a log/log scale for the ordinates and a log scale for the abscissae. The plot for any single liquid will then usually approximate to a straight line, with certain exceptions, the slope of this line being related to the viscosity index. Typical curves are shown in **Fig 2.1.** Where full plots are available for a particular liquid the actual viscosity at any particular temperature can readily be determined.

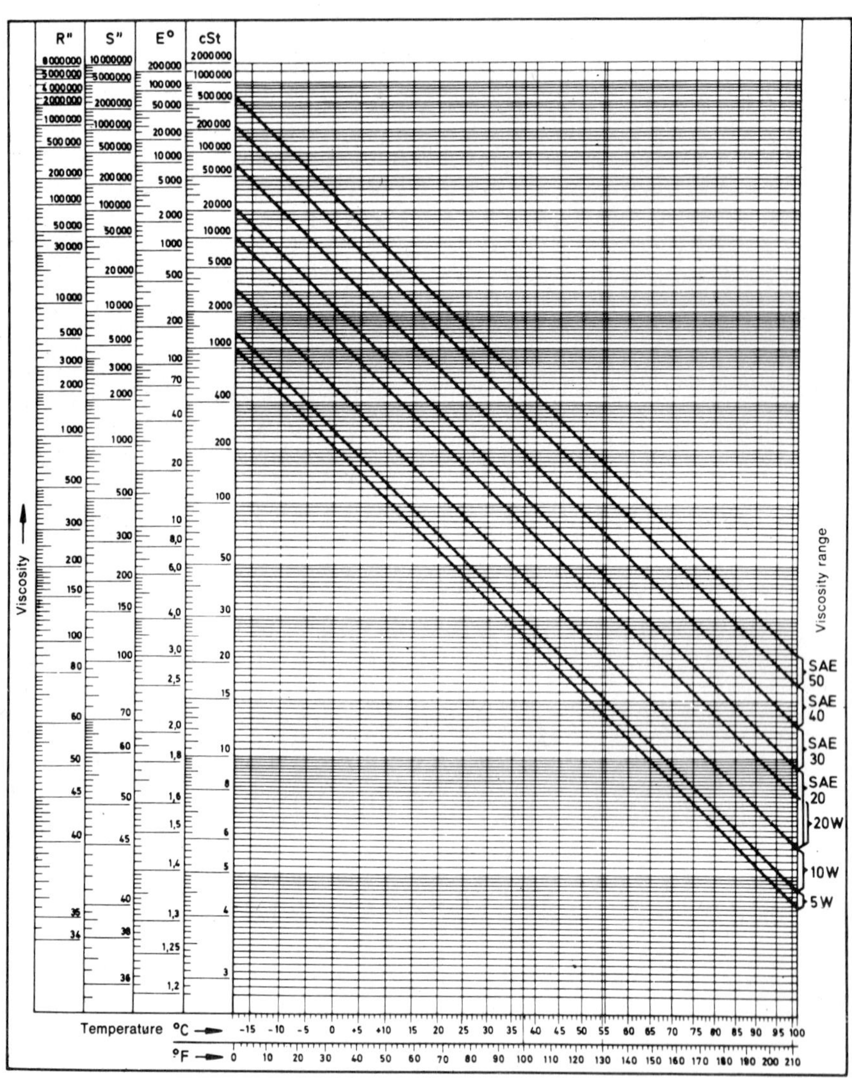

Fig 2.1: Viscosity – temperature chart. (Bosch)

The value of such a chart, however, lies in the 'straight line' function implied. Thus if viscosity figures are known for a particular fluid at two different temperatures only, a line joining these two values will give a fair approximation of the viscosity/temperature characteristics – **Fig 2.2**. Greater accuracy, and a check on the validity of the 'straight line' assumption is, of course, possible if additional viscosity figures are available, particularly to 'justify' extrapolation.

Data may also be available for giving specific viscosity values at different temperatures for different liquids, although in the case of water and most aqueous fluids the effect of temperature can normally be ignored.

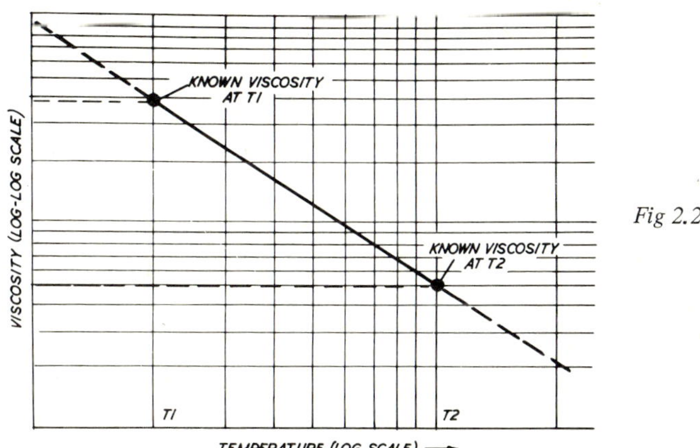

Fig 2.2

Viscosity/Pressure Effects

Viscosity is also affected by pressure, although for most practical applications this effect can be neglected. Basically, fluid viscosity tends to increase with increasing pressure, the effect being most marked at lower temperatures. In the case of water and aqueous fluids, the effect of pressure on viscosity can be neglected entirely. With oils, the effect of pressure can be an increase in viscosity of as much as 30%, per 1 000 lb/in^2 (70 bar) and may have to be taken into account in high pressure systems. Some typical figures are given in **Table VIII**.

Vapour Pressure

The vapour pressure of a liquid is the pressure exerted by the saturated vapour in contact with the liquid surface at any specific temperature, and once that pressure reaches the ambient pressure the liquid will boil. Equally, the vapour pressure at a particular temperature is the absolute pressure at which the liquid would boil at that temperature. This can have a significant effect on pump suction performance in the case of volatile fluids or fluids with high vapour pressure, causing vaporization of the liquid in the suction pipe. This is most readily avoided by providing enough head on the pump suction so that the pressure in the suction

TABLE VIII

VISCOSITY/PRESSURE CHARACTERISTICS FOR MINERAL OILS
(at constant temperature)

Oil Type	Viscosity at Pressure of lb/in^2 (bar) 0 (0)	1 000 (70)	2 000 (140)	3 000 (210)	4 000 (280)	5 000 (390)	6 000 (420)
Naphthenic	100	130	160	200	250	300	350
Paraffins	100	120	130	140	160	200	240
Mixed Base	100	130	160	190	220	270	300
SAE 10 Lube Oil	100	112	137	162	189	222	261
SAE 30 Lube Oil	100	124	157	194	241	275	374

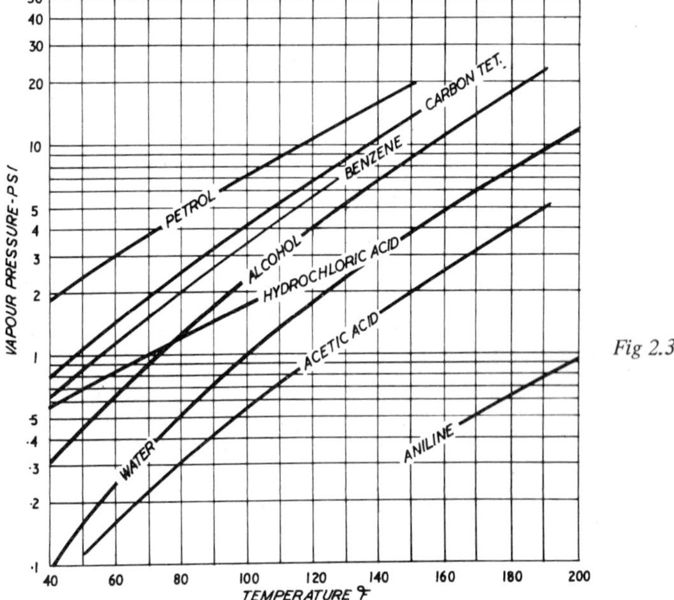

Fig 2.3

pipe is always greater than the vapour pressure of the liquid being inducted. In other words, the vapour pressure of the liquid must be taken into account in arriving at a figure for available net positive suction head (NPSH).

Vapour pressure values for some common liquids are given in **Fig 2.3**. Vapour pressure figures are normally quoted in pressure (*eg* lb/in^2 or bar), but can readily be converted into equivalent head.

3. Fundamentals of Pump Performance

THE MAIN QUANTITIES involved in pump performance are capacity and head, together with input power requirements. *Capacity* is the general term for pump delivery (*ie* volume of fluid handled through the pump in unit time). There is still no complete consistency in capacity units employed but the most common are (Imperial) gal/min or gal/h, US gal/min; lit/min and m^3/h. Larger capacities may also be expressed in units of (English) long tons/h; (American) short tons/h; metric tonnes/h; or (US) barrels/day.

As far as the pump output performance is concerned *head* represents the 'lift' or pressure developed on the fluid. At the same time head (or pressure) is present at different points of the complete pumping system.

Head and pressure are directly related, but *head* is commonly used instead of pressure since it can be expressed by a single dimension (*ie height* in feet or metres) which is independent of the cross section of any liquid column involved. Pressure, on the other hand is force per unit area, but various descriptive terms need understanding in order to avoid confusion. This can be most simply explained by reference to **Fig 3.1**.

Theoretically zero pressure, such as would be attained in a perfect vacuum, is known as *absolute zero pressure.* All practical pressures are then made either relative to this, or to *atmospheric pressure.* Thus *gauge pressure* invariably refers to pressure measurement made with atmospheric pressure as the datum and is either positive or negative (vacuum). The corresponding *absolute pressure*

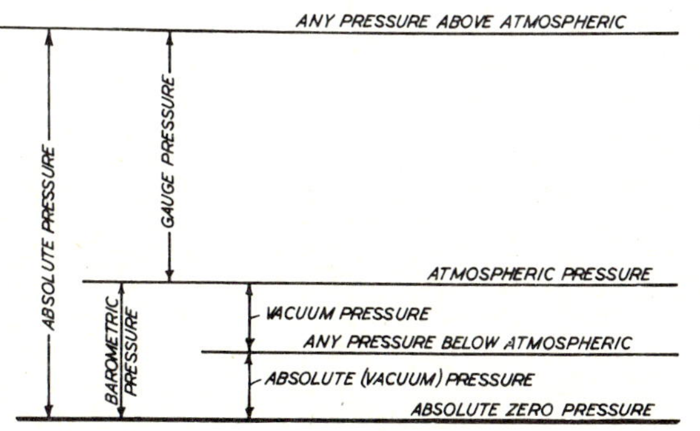

Fig 3.1

then follows from the algebraic sum of the gauge pressure and atmospheric pressure.

In the case of low pressure levels (and this includes vacuum pressures which cannot be greater than 14.4 lb/in^2 (1 bar)), the difference between gauge pressure and absolute pressure can be extremely significant. With high gauge pressures the addition of atmospheric pressure to convert to absolute pressure may be insignificant and in such cases can generally be ignored.

Pressure and Head Relationship

The basic relationship between pressure (P) and head (H) is

$$P = wH \qquad \text{or} \qquad H = P/w$$

where

w is the specific weight of the fluid involved.

In the case of clear, cold water

$P(lb/in^2) = 0.434\ H(feet) \quad H(feet) = 2.31\ P(lb/in^2)$

$P(bar) = 0.1\ H(metres) \quad H(metres) = 10\ P(bar)$

It follows that for any other liquid:

$$P(lb/in^2) = 0.434\ H(feet) \times SG \qquad H(feet) = \frac{2.31\ P(lb/in^2)}{SG}$$

$$P(bar) = 0.1\ H(metres) \times SG \qquad H(metres) = \frac{10\ P(bar)}{SG}$$

where

SG = specific gravity of fluid.

TABLE IA – FEET OF WATER* EQUIVALENT TO PRESSURE (lb/in^2)

Pressure	lb/in^2									
lb/in^2	0	1	2	3	4	5	6	7	8	9
0	–	2.3	4.6	6.9	9.2	11.5	13.8	16.1	18.5	20.8
10	23.1	25.4	27.7	30.0	32.3	34.6	36.9	39.2	41.5	43.8
20	46.1	48.4	50.8	53.1	55.4	57.7	60.0	62.3	64.6	66.9
30	69.2	71.5	73.8	76.1	78.4	80.7	83.0	85.4	87.7	90.0
40	92.3	94.6	96.9	99.2	101.5	103.8	106.1	108.4	110.7	113.0
50	115.3	117.6	120.0	122.3	124.6	126.9	129.2	131.5	133.8	136.1
60	138.4	140.7	143.0	145.3	147.6	149.9	152.2	154.6	156.9	159.2
70	161.5	163.8	166.1	168.4	170.7	173.0	175.3	177.6	180.0	182.2
80	184.5	186.4	189.2	191.5	193.8	196.1	198.4	200.7	203.0	205.3
90	207.6	209.9	212.2	214.5	216.8	219.1	221.5	223.8	226.1	228.4
100	230.7	–	–	–	–	–	–	–	–	–

***Note:** for head in feet for any other liquid divide table figure for pressure by SG of liquid.

TABLE IB – PRESSURE (lb/in^2) EQUIVALENT TO HEAD (Feet of Water)

Head	Feet									
	0	1	2	3	4	5	6	7	8	9
0	–	0.433	0.865	1.300	1.734	2.168	2.601	3.035	3.468	3.902
10	4.335	4.8	5.2	5.6	6.1	6.5	6.9	7.4	7.8	8.2
20	8.670	9.1	9.5	9.8	10.4	10.8	11.3	11.7	12.1	12.6
30	13.005	13.4	13.9	14.3	14.7	15.2	15.6	16.0	16.4	16.9
40	17.340	17.8	18.2	18.6	19.1	19.5	19.9	20.4	20.8	21.2
50	21.675	22.1	22.5	23.0	23.4	23.8	24.3	24.7	25.1	25.6
60	26.010	26.4	26.9	27.3	27.8	28.2	28.6	29.0	29.5	29.9
70	30.346	30.8	31.2	31.6	32.1	32.5	32.9	33.4	33.8	34.2
80	34.681	35.1	35.5	36.0	36.4	36.8	37.3	37.7	38.1	38.6
90	39.016	39.5	39.9	40.3	40.8	41.2	41.6	42.1	42.5	42.9
100	43.349	–	–	–	–	–	–	–	–	–

***Note:** for pressure of a given head of other liquids multiply table figure by SG of liquid

See also **Tables IA** and **IB** for pressure and head conversions. (*Note* tabular data is unwarranted for metric units because of the simple conversions involved).

In pumpwork calculations it is largely immaterial whether head or pressure be used, provided the one unit is used consistently throughout. Strictly speaking, pressure is the real parameter involved, but certain problems are more realistically visualised or presented in terms of equivalent head.

Hydraulic Horsepower

The hydraulic horsepower or equivalent power output of a pump is given by

$$\text{HP (output)} = \frac{\text{Q x H x SG of fluid}}{\text{K}}$$

where

Q is the delivery

H is the total head developed delivery side

K is a constant depending on the units employed

(*eg* for Q in gal/min and H in ft, K = 33)

HP input required and hydraulic horsepower developed are directly related by the efficiency of the pump, *ie*

$$\text{HP input} = \frac{\text{HP (output) x 100}}{\eta}$$

when η is the pump efficiency expressed as a percentage.

It is important to understand which figure is involved where pump performance may be quoted in terms of horsepower or kilowatts (kW) equivalent. Although the efficiency of the pump involved may be high (*eg* 95% or more) in the case of a very large pump the power difference (*ie* actual power absorbed by the pump) may be considerable.

Static Head(s)

In a typical system a pump is called upon to raise liquid from a lower level to a higher level, as in **Fig 3.2**. The vertical lift involved between the lower level and the pump is known as the static suction head. These two vertical distances are always measured from the centreline of the pump, being positive if above the centreline and negative if below. It is important, however, that the static discharge head be measured to the point of free delivery, the exact point of which may not always be easy to determie, although the three most general cases are shown in the diagram.

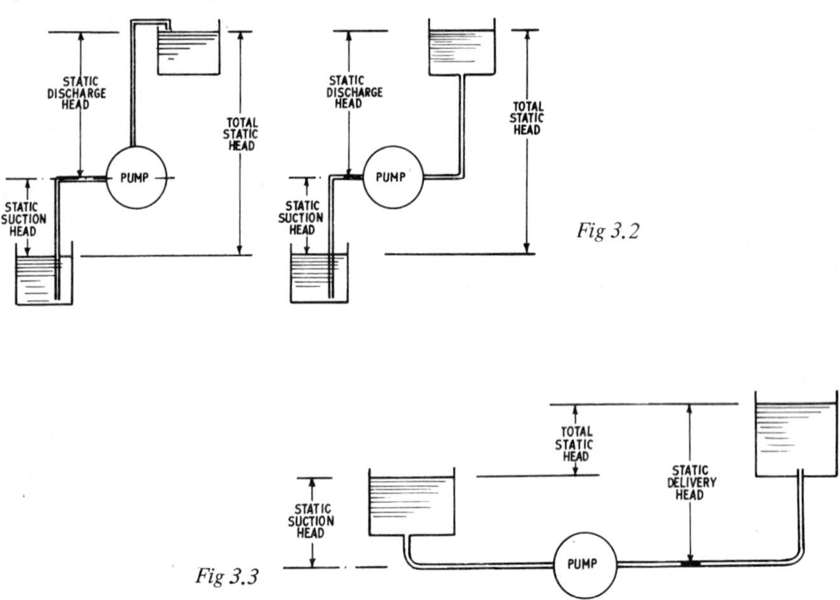

Fig 3.2

Fig 3.3

The total static head involved then follows from subtraction of these two heads, taking into account the signs involved. Thus, where the suction head is positive – **Fig 3.3** – the total static head is less than the discharge head.

To avoid possible confusion it is easier to describe total static head as the difference between the two liquid levels involved (the point of free delivery defining the upper liquid level). It is also helpful to distinguish between a negative suction head (below the centreline of the pump) and a positive suction head, by using the description *(static) suction lift* in the former case.

Note that in computing static heads, horizontal runs are ignored and only vertical heights are concerned, measured relative to the plane of the centreline of the pump.

Suction Lift

A negative static suction head means that the pump is required to lift the fluid through that height to fill the pump before it can start to deliver fluid. To do this the pump must be capable of reducing the pressure at the pump inlet.

Theoretically, by reducing the absolute pressure to zero at the pump inlet the amount of suction lift possible is equal to the head equivalent of the fluid at ambient pressure. Thus in the case of clean cold water at sea level the normal atmospheric pressure is 14.7 lb/in^2 (1 bar) yielding a head equivalent, or theoretical suction lift, of 33.9 ft (10.33 m), commonly rounded off to 34 ft or 10 m.

TABLE II – EFFECT OF ALTITUDE ON THEORETICAL SUCTION LIFT

Altitude		Atmospheric Pressure		Head of Water	
Feet	Metres	lb/in^2	Millibar	Feet	Metres
Sea Level	0	14.7	1 013	33.9	10.33
1 000	305	14.2	979	32.8	10.0
1 500	457	13.9	958	32.1	9.8
2 000	610	13.7	945	31.5	9.6
4 000	1 220	12.7	876	29.2	8.9
6 000	1 830	11.8	814	27.2	8.3
8 000	2 440	10.9	752	25.2	7.7
10 000	3 050	10.1	696	23.4	7.15
15 000	4 570	8.3	572	19.2	5.85

Note: 1 lb/in^2 = 68.95 millibar

This figure will, however, be modified by differences in water temperature affecting its density, and also differences in atmospheric pressure (most significantly, reduction in atmospheric pressure with altitude) – see **Table II**.

In the case of other fluids (or the change in density of water with temperature)

$$\text{Theoretical maximum suction lift} = \frac{33.9}{SG}\ (\text{ft}) = \frac{10.33}{SG}\ (\text{m})$$

where SG is the specific gravity of the fluid at the temperature involved.

Maximum Suction Lift Available (MSLA)

In practice the actual maximum suction lift attainable, or MSLA, is modified by *dynamic losses* or friction losses involved in the suction piping system and the effect of vapour pressure of the fluid at the temperature at which it is being pumped. It is, of course, dependent on the ability of the pump itself to develop

TABLE III – SUCTION CHARACTERISTICS OF PUMPS*

Pump Type	Maximum Suction Lift		Likely Practical Maximum Suction Lift		Remarks
	Feet	Metres	Feet	Metres	
Centrifugal : small					Not self-priming – usually worked with flooded suction
Centrifugal : single-stage	27	8.25	20	6.0	Not self-priming
Centrifugal : two-stage	27	8.25	15	4.5	Not self-priming
Centrifugal : multi-stage	27	8.25	15	4.5	Not self-priming
Centrifugal : self-priming	27	8.25	15	4.5	
Centrifugal : regenerative	27	8.25	15	4.5	Self-priming
Centrifugal : mixed flow	15	4.50	5–10	1.5–3	Not self-priming
Centrifugal : axial flow					Not self-priming – usually operated on low head or flooded suction
Borehole : immersible			Nil	Nil	Flooded suction
Submersible			Nil	Nil	Flooded suction
Reciprocating : direct-acting	27	8.25	22	6.7	Self-priming
Reciprocating : plunger	27	8.25	22	6.7	Self-priming
Reciprocating : piston	27	8.25	22	6.7	Self-priming
Radial piston	27	8.25	22	6.7	Self-priming
Gear			5	1.5	Usually operated with flooded suction
Vane			6	1.8	Can self-prime at high speeds – usually operated with flooded suction

*Extracted from Pumping Manual.

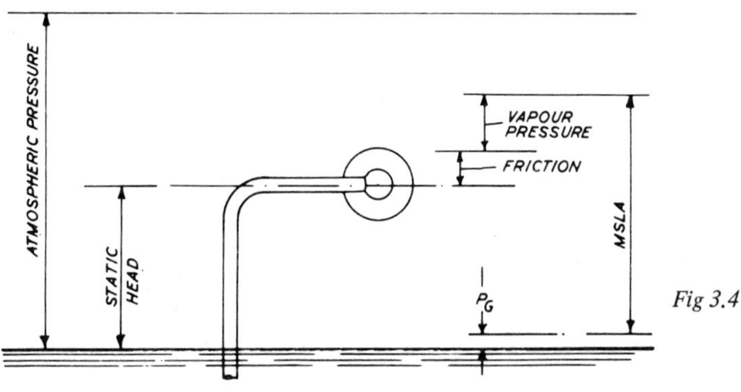

Fig 3.4

pressure reduction (suction) in the inlet pipe. Some types of pumps have little or no ability to do this (see also **Table III**).

Parameters involved are defined in **Fig 3.4** when maximum suction lift available is given by

$$MSLA = f \pm hs + P_V - P_g$$

where

f = the sum of all dynamic and friction losses expressed as a head

hs = static head, either negative if below the pump, or positive if above the pump

P_V = vapour pressure of fluid

Pg = gauge pressure on the surface of the liquid (gauge pressure being the measured pressure above atmosphere)

Net Positive Suction Head (NPSH)

Suction performance is normally expressed in terms of net positive suction head (NPSH) where

NPSH = atmospheric pressure – MSLA

Basically this implies that under given operating conditions (friction losses and static suction lift), the absolute pressure at the pump intake must not be less than the vapour pressure of the liquid being pumped, otherwise the liquid will flash into vapour and suction flow will break down or cause cavitation in the pump.

For any given conditions the *available* NPSH can readily be calculated on the lines above. As far as a pump is concerned, it will require a certain NPSH or pressure available to perform satisfactorily, this *required* NPSH being a function of the pump design, and also varying with the capacity and speed of the pump (particularly in the case of impeller pumps).

In simple terms, the required NPSH of a pump is the 'allowance' necessary to cover internal head losses within the pump. Part of these losses will be produced by friction and viscous drag, part by internal leakage (in the case of positive displacement pumps), and part by any dynamic depression head generated within the pump (*eg* with rotodynamic pumps).

Given a required NPSH for a pump to operate at rated capacity, the *allowable* suction lift can be determined by deducting suction friction head and vapour pressure from the theoretical suction lift available (suitably corrected for atmospheric pressure or altitude, and temperature). If the sum to be deducted from the theoretical suction lift is greater, then the difference is the *positive* static head necessary to ensure satisfactory operation.

The required NPSH of a pump is a manufacturer's figure, and specifically related to speed and capacity in the case of rotodynamic pumps. It also varies widely with different types of pumps – *eg* from virtually zero in the case of reciprocating pumps to a high proportion of the theoretical suction head with axial flow pumps (which are normally operated on flooded suction). Figures may be quoted against water temperature, or against capacity (and in the latter case often allied to suction lift possible at such capacities). It is a general rule with all rotodynamic pumps that as the available NPSH decreases the capacity also decreases. Thus for given operating conditions the higher the suction lift required the lower the capacity for a given size of pump, and vice versa. Excessive suction lifts, or insufficient NPSH, generally leads to cavitation. This condition (or more exactly a reduction in available NPSH) can also result from a rise in fluid temperature increasing the vapour pressure.

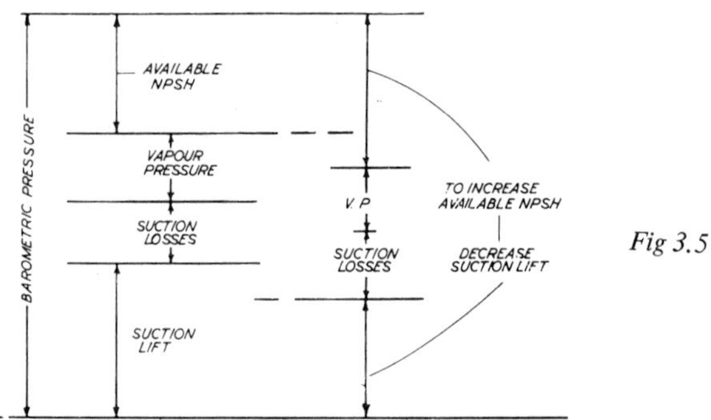

Fig 3.5

The required NPSH can be represented diagrammatically, as in **Fig 3.5**, which illustrates the importance of vapour pressure considerations. Vapour pressure is constant for a given liquid at a given temperature. The required NPSH is a function of the pump design and so, since all other factors are known using such a diagram, the permissible suction lift can be calculated. Working over a greater lift than this would almost certainly cause problems by fluid vaporization.

Where the suction supply is from a closed tank or sealed reservoir, tank pressure is substituted for atmospheric pressure, and converted to equivalent head to be in consistent units. NPSH, in fact, represents the pressure-head required or available to force a given flow rate of specified fluid through the suction piping into the pump, but where it is given on pump characteristic curves it is normally presented as a head figure and corrected to the centreline of the pump.

Suction Systems and NPSH Available

The general case is where the liquid level is under atmospheric pressure and the effect of fluid vapour pressure can be ignored (*eg* clean, cold water); with either a positive suction head (+hs) or suction lift (–hs) – **Fig 3.6**.

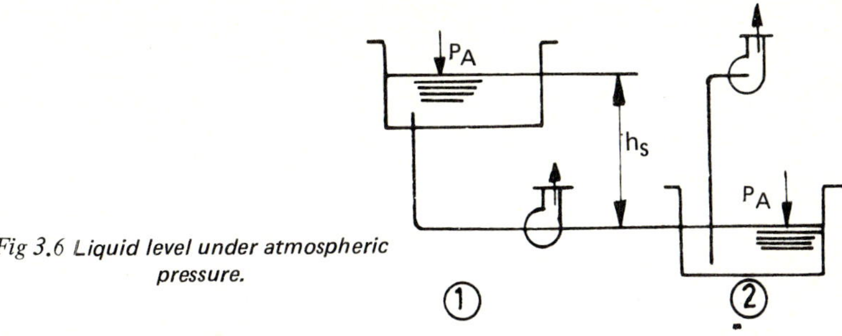

Fig 3.6 ***Liquid level under atmospheric pressure.***

The following formulas then give NPSH available:

$$\text{NPSH available (English units)} = \frac{1.13}{\delta}(P_A - 2.04\,P_v) \pm h - h_f$$

$$\text{NPSH available} = \frac{10.2}{\delta}(P_A - P_v) \pm h - h_f$$

The second case, shown in **Fig 3.7**, is where the liquid level is exposed to pressure P_1 when

$$\text{NPSH available} = \frac{2.31}{\delta}(P_1 + 0.49\,P_A - P_v) \pm h_s - h_f$$

$$\text{NPSH available} = \frac{10.2}{\delta}(P_1 + P_A - P_v) \pm h_s - h_f$$

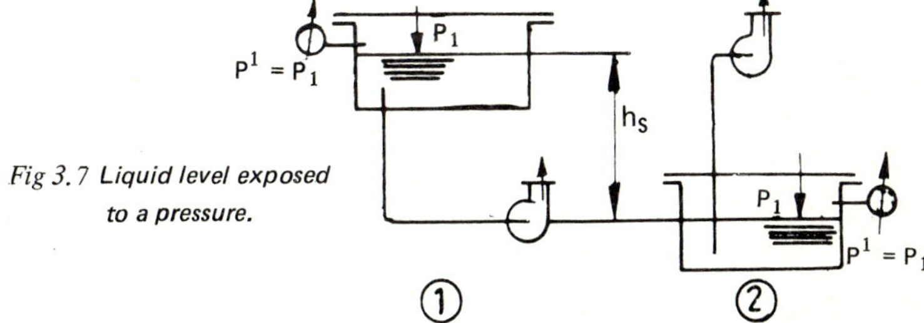

Fig 3.7 ***Liquid level exposed to a pressure.***

In the third case, shown in **Fig 3.8**, the liquid level is under vapour pressure of fluid, when

$P_1 = P_v - P_A = P_v - 0.00134\,P_A$ (metric units)

$P_1 = P_v - 0.49\,P_A$ (English units)

These values are used in the previous formula.

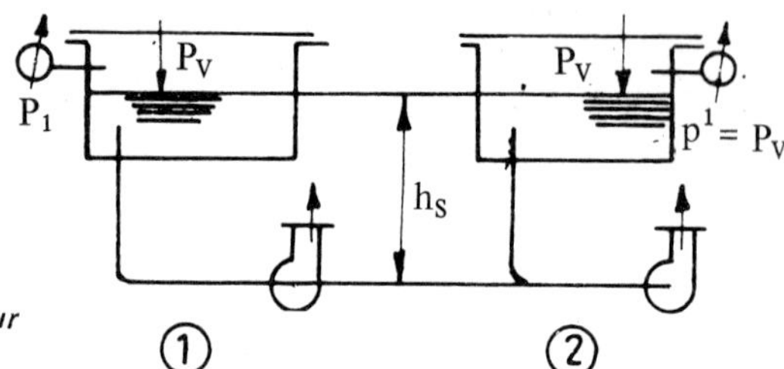

Fig 3.8 ***Liquid level under vapour pressure of liquid.***

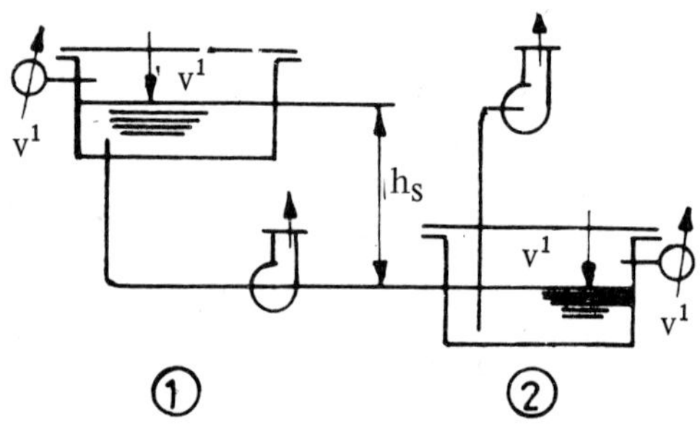

Fig 3.9 ***Liquid level exposed to a vacuum.***

The final case is where the liquid level is exposed to a vacuum (vacuum pressure v^1) – **Fig 3.9.** Formulas for calculating the NPSH available are then:

NPSH available (English units) $= \dfrac{1.13}{\delta}(P_A - V - 1.04\,v^1) \pm h - h_f$

NPSH available $= \dfrac{10.2}{\delta}\left(P_A - \dfrac{v^1}{1\,000} - P_V\right) \pm h - h_f$

Pump Output Requirements

On the delivery side, a pump must develop sufficient head to meet the total head requirements which must take into account:

(i) Any difference between liquid levels on inlet and outlet sides;

(ii) Any difference between vacuum and/or pressure on the inlet and outlet side liquid free surfaces;

(iii) Frictional resistance, *ie* head required to overcome resistance to flow through pipes and fittings;

(iv) The velocity head at point of final discharge.

Dynamic heads must be considered separate from static heads and their contribution can vary widely with different installations. The total system head will govern the working part of the pump in the case of rotodynamic pumps, or the pressure loading in the case of positive displacement pumps.

Position Head

Whilst *pressure head* refers specifically to the height of a column of liquid, *position head* refers to the height of a specific point above a chosen datum level. Thus in the elementary system shown in **Fig 3.10** the pressure head is defined by the difference in fluid levels whilst position head may relate to any chosen point in the system, and to any chosen datum plane.

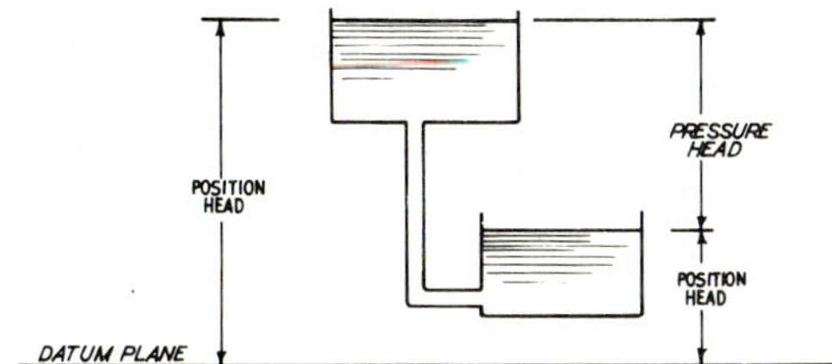

Fig 3.10

Velocity Head

The fact that a fluid is flowing with a velocity V implies either expenditure of energy to achieve that velocity, or change of energy if that velocity is modified or dissipated (*eg* by the fluid flowing into an outlet tank). The energy related to velocity is best described in terms of *velocity head,* which is the height through which an object would have to fall in order to attain that velocity.

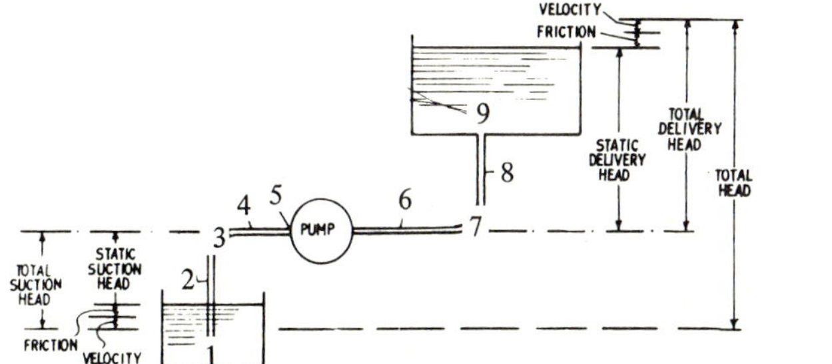

Fig 3.11

The System Analysed

In any practical system energy losses are involved by friction and modifications of velocity head, which have the effect of modifying the static heads involved. Thus, considering the simple system shown in **Fig 3.11**, losses are involved at the various numerical points indicated, *viz:*

point 1 – a loss of velocity head is experienced at the inlet which may be equal to the full theoretical value of the velocity head, or reduced below this value by suitable shaping.

point 2 – frictional loss in straight run of pipe.
point 3 – friction loss in bend.
point 4 – frictional loss in straight run of pipe.
point 5 – frictional loss in suction fitting.

Items 2, 3, 4 and 5 can be summed as a complete frictional loss on the suction side, or *suction friction head.* The *effective suction lift* is then the sum of static suction lift, velocity head loss, and friction head.

On the delivery side, points 6, 7 and 8 represent pipe and fitting losses, again summable as a *discharge friction head;* and point 9 an exit or outlet loss, expressed in terms of velocity head loss. Thus the *effective discharge head* becomes the sum of the static discharge head, discharge friction head and velocity head.

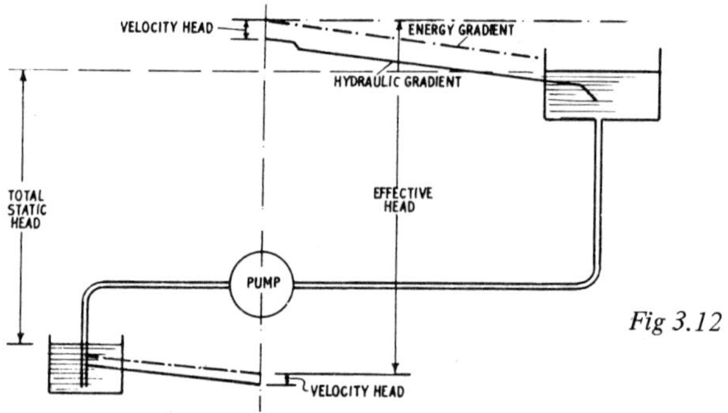

Fig 3.12

Hydraulic Gradient

Fig 3.12 shows the same system as **Fig 3.11** annotated with reference to the pressure heads and energy levels throughout the system. The respective traces are known as the hydraulic gradient line and energy gradient line, the vertical distance between these lines at any point being the velocity head at that point. Both lines drop in the direction of flow and for a constant area pipe this slope is regular and equal to the head loss due to friction. The term *hydraulic gradient* is used to define the slope of the hydraulic grade line and is numerically equal to the head loss per unit length.

Manometric Head

Fig 3.13 shows the system of **Fig 3.11** annotated with reference to direct pressure measurement applied to the system, as when measuring actual pump performance. Pressure gauges are connected to the suction and discharge sides of the pump to read directly the pressure difference generated by the pump. These gauges should be mounted on the same horizontal plane as the pump centreline, otherwise corrections are needed to the readings obtained in order to eliminate gauge heads introduced.

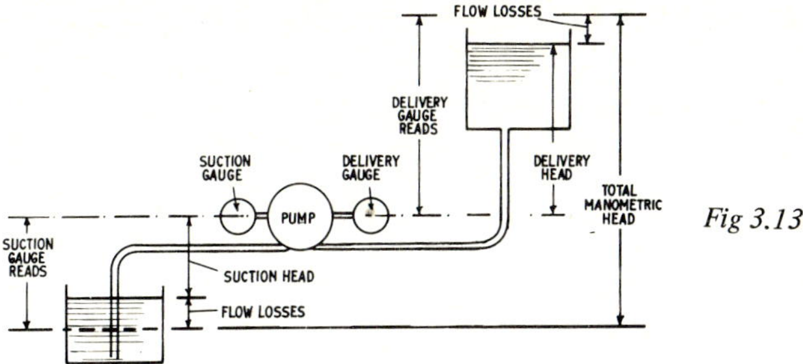

Fig 3.13

Pressure readings are read or converted into head of liquid being pumped, when

$$\text{Effective Head } (H_e) = H_s + \frac{(V_d)^2}{2g} + H_d - \frac{(V_s)^2}{2g}$$

where

H_s = head as read on suction gauge
H_d = head as read on delivery gauge
V_s = mean flow velocity at suction flange
V_d = mean flow velocity at delivery flange

The manometric head (H_m) is equal to the sum of H_s and H_d and is a direct measure of the pressure head generated by the pump. The effective head, on the other hand, is the increase of total energy generated by the pump.

4. Centrifugal Pumps

THE COMPLETE GROUP comprising centrifugal, mixed flow and axial flow pumps are also described as *impeller pumps,* but this is open to objection on the score that there are other forms of pumps with impellers which have positive displacement characteristics (and thus classified as positive rotary pumps). More acceptable group descriptions are *rotodynamic pumps* (in preference to rotokinetic pumps, which means the same thing); *Kreiselpumpen* (used almost exclusively in Germany), and *turbo pompes* (in general use in France). A further alternative sometimes quoted is *turbo machines.*

The description *centrifugal* pump is generally given to machines employing an impeller with fixed blades housed in a suitably shaped casing, the impeller being mounted on a rotating shaft and the casing fixed. Pumping action, or liquid transport from the inlet to the outlet side, is achieved by virtue of an increase in momentum applied to the liquid. At the same time the resulting fluid motion through the pump produces a reduction in pressure on the inlet side, or suction effect. Such machines form a complete class of pumps, with behaviour not immediately determinable because the throughflow lacks positive displacement characteristics. In other words, the two main characteristics – quantity and delivery pressure or head – are interdependent and related to impeller shape, as well as pump size and speed.

Typical small centrifugal pump with foot mount and close coupled motor.

The impeller form, and the shape of its attendant casing, also determines the manner in which the liquid is accelerated through the pump. The true centrifugal pump utilises an impeller comprising a series of blades set between two discs giving a *radial* velocity of flow through the space between the discs, *ie* the liquid is thrown outwards radially into a volute shaped casing. Other impeller forms, however, may direct flow in both an axial and a radial direction; or in a purely axial direction. These are known generally as *mixed flow* and *axial flow* pumps, respectively.

Possible confusion can be avoided by adopting the main classifications centrifugal, mixed flow and axial flow, although mixed flow pumps admit of further subdivision. Thus where both radial and axial flow is derived from an impeller having blades with inlet and outlet edges inclined to the axis delivering into a volute casing, the type may be called a *helical* (or *helicoidal)* pump. Where the impeller form is similar but followed by guide vanes in a bowl shaped casing, the type is often called a *bowl* pump, or sometimes a *diagonal* pump. Pure axial flow pumps are commonly called *propeller* pumps.

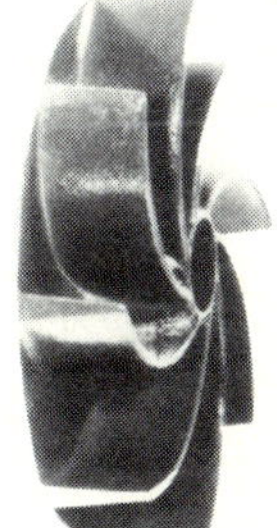

Radial flow impeller (left), single-vane impeller (centre) and free flow impeller (right).

Diffusers

Centrifugal pumps which have stationary guide vanes surrounding the impeller are generally known as diffuser-type pumps. The purpose of these vanes is to provide gradually divergent passages to change the direction of liquid flow and convert velocity energy into pressure head, with better recovery than that provided by a simple expanding spiral casing under certain conditions (*eg* at low specific speeds).

It should be noted, however, that a volute itself is a recuperator and a correctly designed volute will produce small hydraulic losses, whilst further recuperation can take place in a straight diffuser in which the volute terminates. Thus the efficiency of a volute type pump may be higher than that of a diffuser-type pump without the attendant disadvantages of increased bulk, weight and cost.

Regenerative Pumps

A further type which must be included is the *peripheral* or *regenerative* pump which employs a distinctive impeller form like a turbine wheel with the blades running in an annular channel surrounding the periphery of the wheel. Pumping action is provided by the series of impulses given to the fluid by the rotating blades.

The true peripheral pump has a double-sided channel in which the liquid circulates, situated partly in the cylindrical part of the casing and partly in the side plates. The suction space is separated from the delivery space by a barrier on the casing. Peripheral pumps normally have relatively low specific speeds, although shaft speeds may be as high as 6 000 rev/min. Efficiency is relatively low – *eg* 45–50%, and the main advantage of the peripheral pump is its ability to attain a high head, up to 600–650 ft (180–200 m) per stage, and compact size and weight for given operating conditions. They are competitive with, and often replace, multi-stage centrifugal pumps for high head duties with deliveries up to 250 gal/min (1 100 lit/min).

Side-Channel Pumps

A variation is the *side-channel* pump in which the channels adjacent to the impeller blades are cut in the side plates only. By suitable profiling the pump can be made to effectively expel air and thus have self-priming characteristics, which the true peripheral pump lacks (unless coupled to a reservoir and condenser). Again, the side-channel pump is characterised by the ability to develop a very high head per impeller (stage), although their efficiency is only about one half that of the true peripheral pump.

The internal flow generated by regenerative pumps is highly turbulent which makes the wetted components particularly susceptible to erosion-corrosion and wear. Suitable materials of construction must therefore be chosen accordingly.

Selection by General Characteristics

In general, centrifugal pumps are suitable for both low and high heads and may be categorised on this basis, *viz* –

Low head types – maximum total head up to 60 ft (20 m)
Medium-head types – maximum total head 50–200 ft (15–60 m)
High-head types – total head over 200 ft (over 60 m).

The basic design lends itself to generating heads of up to 500 ft (150 m) or more in a single stage; and if still higher heads are required two or more stages can be combined in series to produce a multi-stage pump.

High suction lifts are also possible with centrifugal pumps, provided the specific speed is kept below about 2 500 ft (900 m).

At the other extreme, the axial flow pump is capable of realising very high efficiencies when handling large deliveries at low heads, the relatively high operating speeds obtainable permitting the selection of a much smaller and cheaper machine than would be required in the case of a centrifugal pump. It also maintains good efficiency at a constant speed over a considerably greater range of head than the centrifugal pump (see **Fig 4.1**) and is thus particularly applicable to variable head duties (*eg* land drainage and irrigation services). Suction lift attainable, however, is very low or negligible.

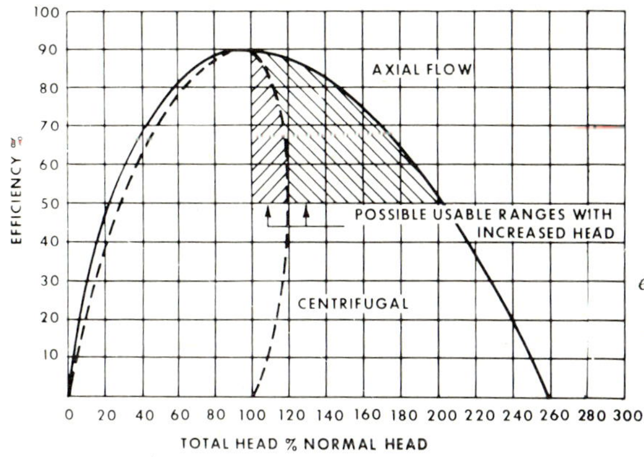

Fig 4.1: Comparison of efficiencies of axial-flow and centrifugal pumps with varying head (constant speed).

Mixed flow pumps have general characteristics between these two, those with lower specific speeds being more similar to centrifugal pumps and those with higher specific speeds more similar to propeller pumps. Their chief application is where a high delivery is required but the total head required is too high to be developed by a pure axial flow pump; and a centrifugal pump giving the required delivery and head would be unnecessarily large and slower running (to operate within the optimum specific speed range) and thus require a larger driver.

A broad guide as to the selection of the most suitable type of rotodynamic pump for particular requirements is given in **Table I.**

Performance Curve Characteristics

More specific information on the differences between centrifugal (or radial flow), mixed-flow and axial-flow pumps can be gleaned from their characteristic performance curves. Typical generalised curves are shown in **Figs 4.2–4.7.** Such curves are normally plotted for a constant speed (design operating speed), when capacity can vary from zero to a maximum, depending on the head. This should not be read as the usable range of the pump, however. Specifically the pump will have maximum efficiency at one point on the H-Q curve only, which is known as the design operating point. Ideally the pump should be operated at this point. Operation at any other point on the H-Q curve will result in a marked loss of efficiency. The effect of variable speed must be considered separately – see later.

TABLE I – BROAD SELECTION GUIDE TO CENTRIFUGAL PUMPS

Specific Speed English (Metric)	Typical Construction	Initial Cost	Efficiency	Suction Lift	H-Q Characteristics	Over-loading	Stable Characteristics	Head Capability	Bulk	Base of Disassembly	Choking
700–2500 (245–875)	Single-suction radial-flow, volute	4	3-5	5	1-3	1-3	0-3	4	2-3	3-4	1-2
700–2000 (245–700)	Single-suction radial-flow, guide ring	1	3-4	4	1-2	1-2	0-2	5	2	1	0
1250–4500 (425–1500)	Double-suction radial-flow, single stage volute	3	4-5	3-4 4	3-4	3-4	1-3	4	3	5	3-4
1250–2500 (425–875)	Double-suction multi-stage, radial-flow, volute	2	4-5	4	3-4	3-4	1-2	5	2	4	3
3000–6000 (1000–2000)	Single-suction, single-stage, mixed-flow, volute	2	5	2	4	4	4	2	3	4	4
3000–6000 (1000–2000)	Single-suction, single-stage, mixed-flow, concentric diffuser	3	4	2	4	4	4	3	4	2	2
5000–7500 (1750–2700)	Double-suction, single-stage, mixed-flow, volute	2	4	2	4	4	4	1	3	4	3
7000–12500 (2500–4375)	Single-suction, single-stage, axial-flow, concentric diffuser	4-5	3-4	0	5	5	5	0	5	4-5	3

Comparative ratings given from 0-5 (the higher the figure the better)

Stable and Unstable H-Q Curves

In the case of centrifugal pumps **(Fig 4.2)** the H-Q curve may be stable or unstable. In the former case maximum H is achieved at zero discharge and the value of H progressively decreases as the discharge rate increases. With unstable characteristics the value of H first increases with increasing discharge to a maximum, when it falls again with further increase in discharge rate.

The effect on power characteristics is also shown in **Fig 4.3**. With a stable H-Q curve the power required rises to a peak at or near the design operating point and then falls again. With an unstable H-Q curve the power required continues to rise after reaching the design operating point. These are known as non-overloading and overloading characteristics, respectively. The difference is that with non-overloading characteristics there is no possibility of overloading the pump driver should the head fall below the design operating point level. With overloading characteristics, should the head fall either the driver will be overloaded, or larger power will be delivered to the fluid (if the driver has a sufficient

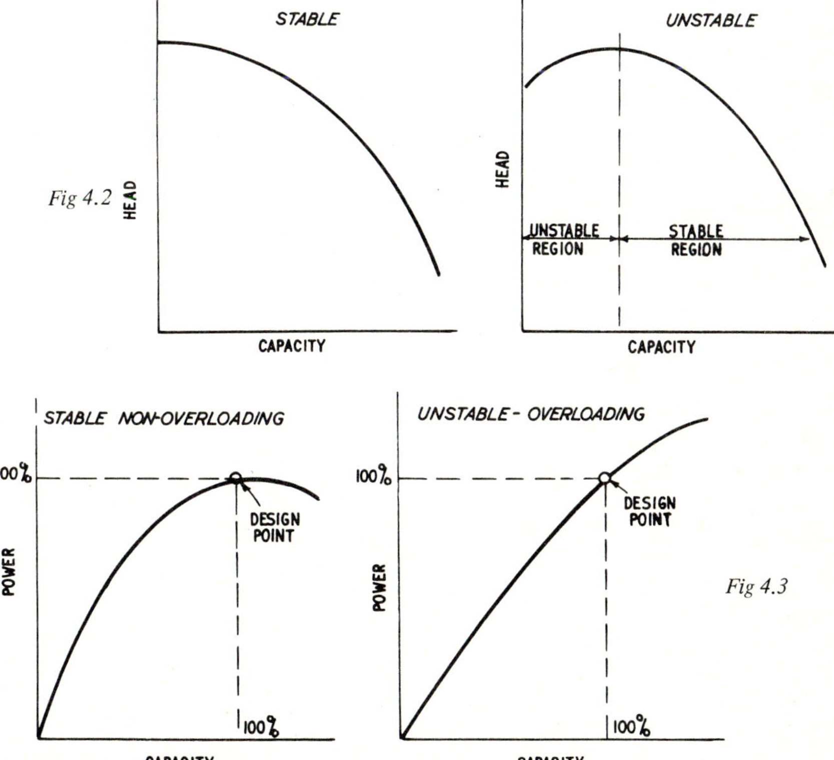

Fig 4.2

Fig 4.3

reserve of power). This could lead to potentially dangerous conditions, such as vibration in the pipeline, which may preclude the use of pumps with unstable H-Q characteristics for specific applications.

Shape of the H-Q Curve

The shape of the H-Q curve is dependent on specific speed, the form and number of blades on the impeller, and the form of the casing and recuperators employed. Overloading is most likely to occur at low specific speeds, where the H-Q curve tends to become flattest, but in the majority of cases it is the pump designer's aim to produce stable (or non-overloading) characteristics.

With increasing specific speed the slope of the H-Q curve becomes steeper and the H-Q characteristics of higher specific speed centrifugal pumps and all mixed-flow and axial-flow pumps are invariable stable and non-overloading. Within the centrifugal pump range, however, the slope of the H-Q curve may range from flat to quite steep, which factor can influence selection.

Curves for two pumps having the same design operating point but quite different H-Q slopes are shown in **Fig 4.4.** The main differences in operating characteristics can be summarised as –

(i) The pump with a flat H-Q curve will show little difference in total head over quite large variations in discharge rate.

(ii) The pump, with the steep H-Q curve will show little difference in discharge over considerable variations in total head.

Type (i) would, therefore, be preferred where the demand is variable but a constant head is required; and type (ii) where there may be considerable variations in total head but a substantially constant discharge is required.

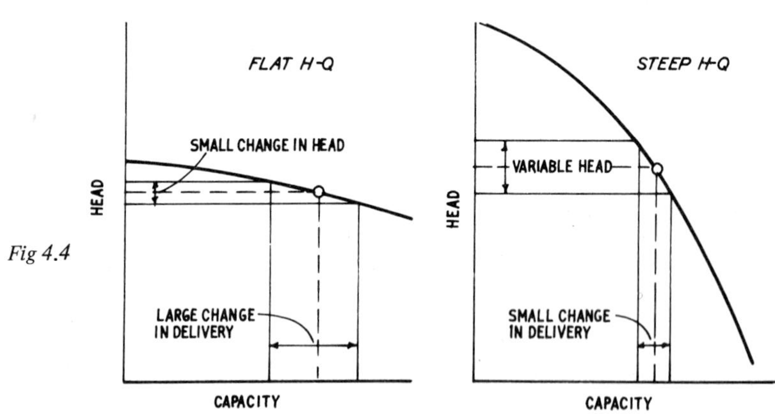

Fig 4.4

Mixed-Flow Pumps

The H-Q curve for a mixed-flow pump is steep with the point of maximum efficiency displaced towards maximum capacity – **Fig 4.5**. The BHP curve, on the other hand, is much flatter and approaches the horizontal, so that power demand will vary little regardless of the working point.

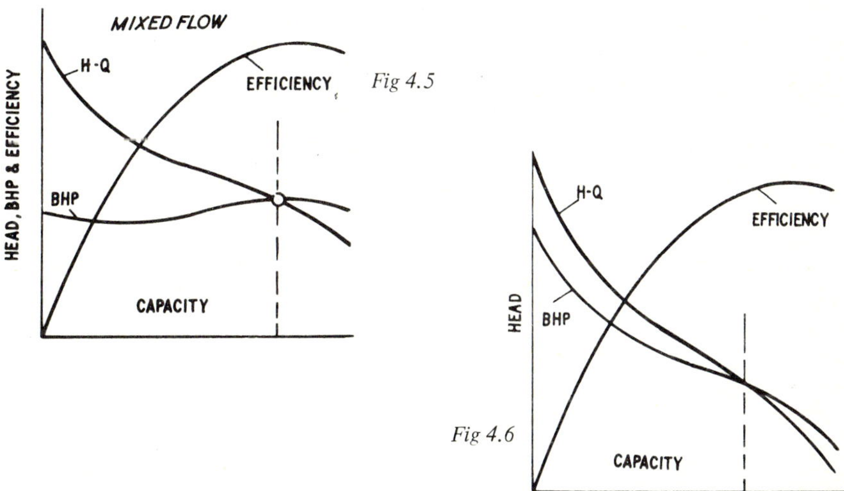

Fig 4.5

Fig 4.6

Axial-Flow Pumps

The H-Q curve for an axial-flow pump is much steeper again – **Fig 4.6**. The power input curve takes a quite different form with maximum power demanded at zero delivery, and a steep fall in power beyond the design working point. High efficiency, however, is maintained over a greater range of head than a centrifugal pump, although the actual attainable head is much lower.

At low outputs the power absorbed by an axial-flow pump rises steeply and at zero discharge may be as high as twice that required for working at the design point. Where an axial-flow pump may be called upon to deliver a reduced output, or to operate under conditions involving a closed discharge, it thus becomes necessary either to provide increased power or a means of unloading the pump when the head rises to a predetermined figure.

Regenerative Pumps

Regenerative pumps normally have straight line H-Q curves with a fairly steep gradient – **Fig 4.7**. Since the head or pressure developed can rise sharply with decreasing capacity they are not suitable for discharge regulation by throttling; or for applications where the discharge may be shut off unless the pump is protected by a suitable pressure relief valve.

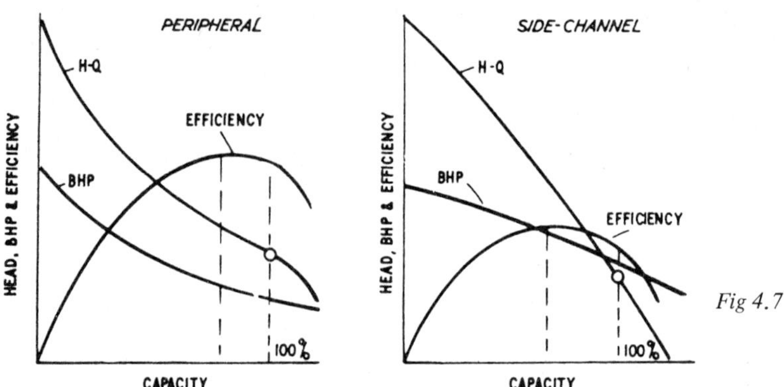

Fig 4.7

The H-Q curve of a centrifugal pump is never complete. That is to say the curve ceases or reaches a cut-off point at a certain value of Q and at this point cavitation is fully developed in the pump. The cut-off point may be quite distinct in the case of a centrifugal pump with a flat H-Q curve. With pumps having higher specific speeds the H-Q curve slopes more gradually to the cut-off point until, in the case of axial-flow pumps, the cut-off point may not be apparent at all.

The cut-off point is also modified by the speed and suction lift. Thus **Fig 4.8** shows five H-Q curves for different speeds for the same low specific speed pump having a sharply defined cut-off. The cut-off in this case occurs at a lower value of discharge with both decreasing speed and decreasing head.

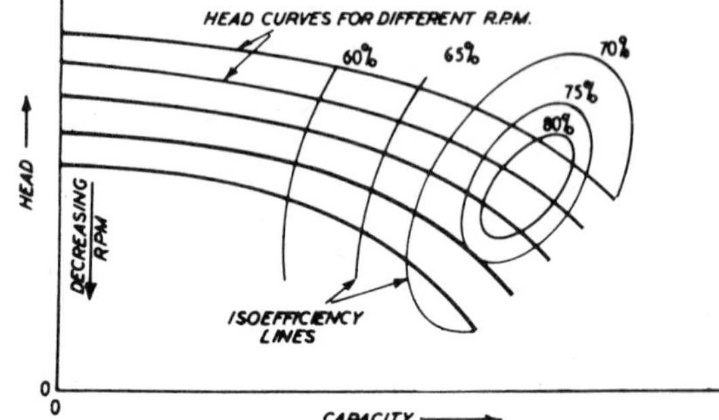

Fig 4.8: Speed and isoefficiency curves, showing also different cut-off points.

Specific Speed

The parameter *specific speed* is commonly quoted for rotodynamic pumps. Basically this represents the speed in rev/min of a geometrically similar pump but of such a size to generate one water horsepower when working against unit head. In practice it defines the optimum rotor geometry for any size of pump, *ie* the optimum specific speed for maximum efficiency although actual efficiency achieved is also affected by capacity. In other words, as a general guide, specific speed can be taken as a suitable operating range for a particular impeller form – see **Figs 4.9** and **4.10**.

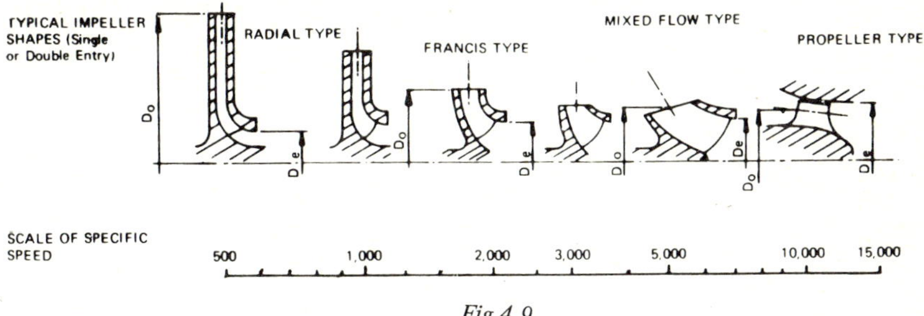

Fig 4.9

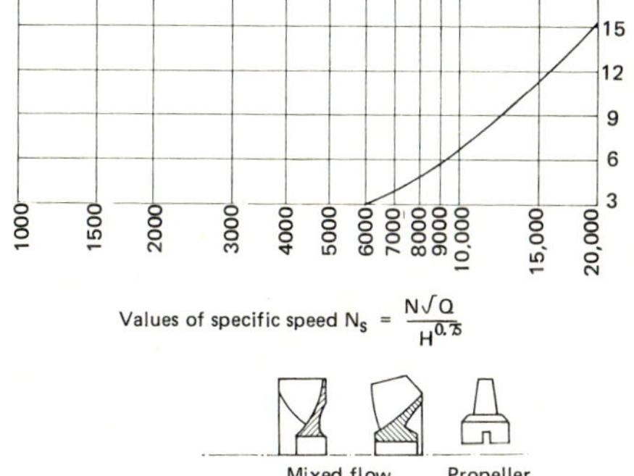

Fig 4.10: Approximate relative impeller shapes and NPSH variations with (metric) specific speed at 4.5 m head.

Specific speed (N_S) is a numerical value calculated on the basis

$$N_s = \frac{\text{rev/min} \times \sqrt{Q}}{H^{0.75}}$$

where

Q is the discharge at the pump design point

H is the effective head at the pump design point

Confusion can arise because of the different units which may be used for H and Q. Head in feet is standard in English and American practice; and head in metres in metric practice. But specific speed calculated in terms of the English (Imperial) gallon for Q will differ from that based on the US gallon, *ie:*

N_s (English) = N_s (American) x 0.76
N_s (American) = N_s (English) x 1.315

Metric specific speed values for the same pump will be different again, further complicated by the fact that two different units may be used for Q, *ie* m^3/sec or m^3/hr. Also the SI unit for rotational speed is radians/sec and not rev/min. The *preferred* metric units are metres (H), m^3/hr (Q) and rev/min,

when

N_s (metric)	=	N_s (English)	x	0.35
	=	N_s (American)	x	0.46
N_s (English)	=	N_s (metric)	x	2.885
N_s (American)	=	N_s (metric)	x	2.20

Sometimes, too, a *nominal* specific speed value is quoted for pumps rather than a true specific speed, based on the simpler equation

$$N_s \text{ (nominal)} = \frac{\sqrt{Q \times \text{rev/min}}}{H^{0.75}}$$

Again, numerical values will vary with the units used.

Specific speed figures can also vary with the method of assessing double-entry and multi-stage pumps. With double-entry impellers some authorities calculate specific speed on total flow through the pump (*ie* Q = rated output); others may use a figure of Q/2 in calculation, giving a difference of 1.414 times between the two values for an identical pump.

In the case of multi-rotor machines, specific speed is correctly defined by the individual performance of each rotor when the rotors are working within the same casing and are carried on a common shaft, whence the correct figure for Q is total flow divided by the number of rotors.

In the case of multi-*stage* machines, where the rotors are working in series, the Q value is equal to the total flow but the true head is the head per stage, *ie* total head divided by a number of stages.

Specific Speed and Efficiency

A general relationship applies between (true) specific speed and efficiency of rotodynamic pumps. The effect of pump size (capacity) is also significant. Optimum specific speed ranges for different impeller forms (*ie* types of rotodynamic pumps), or the range giving competitive efficiency with other pump types is then broadly as follows:-

	Specific Speed		
	English	US	Metric (preferred)
Centrifugal pumps	500 – 2 500	650– 3 250	175– 875
Mixed flow pumps	2 500 – 7 500	3 250–10 000	875– 2 625
Axial flow pumps	7 500 – 15 000	10 000–20 000	2 625– 5 250

Note principally that centrifugal pumps with specific speeds lower than about 500 (175 metric) have increasing non-competitive efficiency.

Specific Speed and Suction Lift

Fig 4.11 illustrates in some detail the likely suction lift capabilities of single-suction centrifugal pumps over a wide range of specific speeds. Note in particular that the suction lift likely to be attainable is related to the total head on the

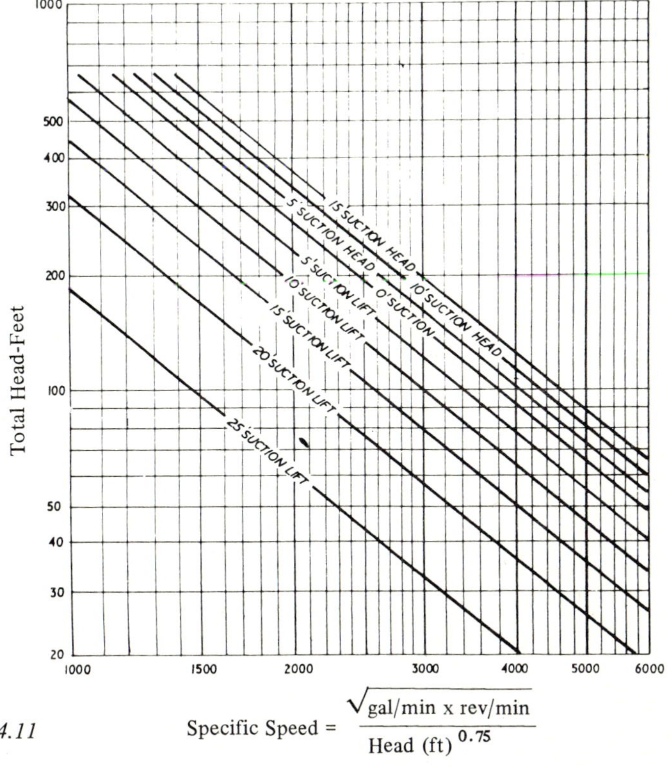

Fig 4.11 Specific Speed = $\dfrac{\sqrt{\text{gal/min x rev/min}}}{\text{Head (ft)}^{0.75}}$

pump, whilst the corresponding specific speed figures are nominally upper limits. This diagram gives a general picture of the suction capabilities of a wide range of rotodynamic pumps from radial flow (true centrifugal) to axial flow types. For completeness the curves have been extended above the zero suction lift line to give positive suction head recommendations as well.

Maximum Suction Lift of Centrifugal Pumps

Theoretically at least the maximum suction lift available with a centrifugal pump can be calculated if the *cavitation factor* (σ) is known, *viz*

$$\text{maximum } h_s = h_b - h_f - h_{vp} - \sigma H$$

where

h_s = maximum suction lift available
h_b = barometer pressure expressed as head
h_f = total frictional and energy losses in suction line expressed as head
h_{vp} = fluid vapour pressure, expressed as head
H = total head developed by the pump

The limitation of this formula is the difficulty in establishing a true value for the cavitation factor. However, a likely value related to the specific speed of the pump can be established from **Fig 4.12.** Normally the upper line value would be taken for a conservative estimate of suction head likely to be available. However, estimated values of the cavitation factor should be used with reservation since other factors may affect the true value. Thus the value of σ tends to decrease somewhat with increasing pump speed rev/min as evidenced by the fact that pumps with slower rotational speeds tend to be more prone to cavitation. Also

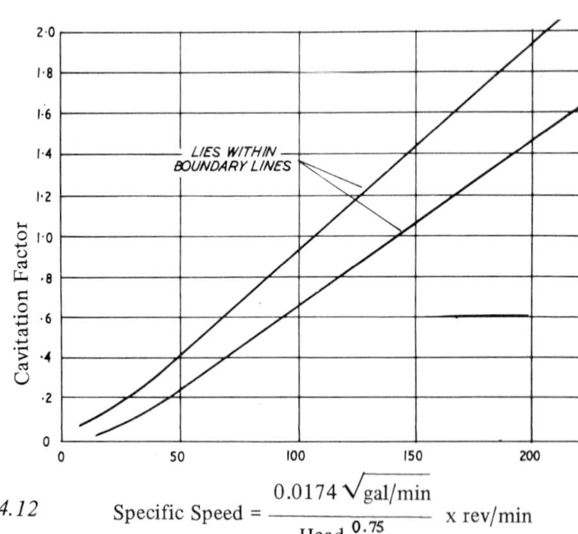

Fig 4.12 Specific Speed = $\dfrac{0.0174\sqrt{\text{gal/min}}}{\text{Head}^{0.75}}$ x rev/min

whilst the cavitation factor increases with specific speed, pumps with higher specific speeds are less prone to cavitation because of the shift of the cut-off point to the right.

Specific Speed and H-Q Characteristics

Fig 4.13 illustrates the effect of specific speed on rotodynamic pump characteristics, *ie* form of H-Q curve and comparative level of maximum head developed, efficiency and input power absorbed; also the position of the pump design operating point (for maximum efficiency).

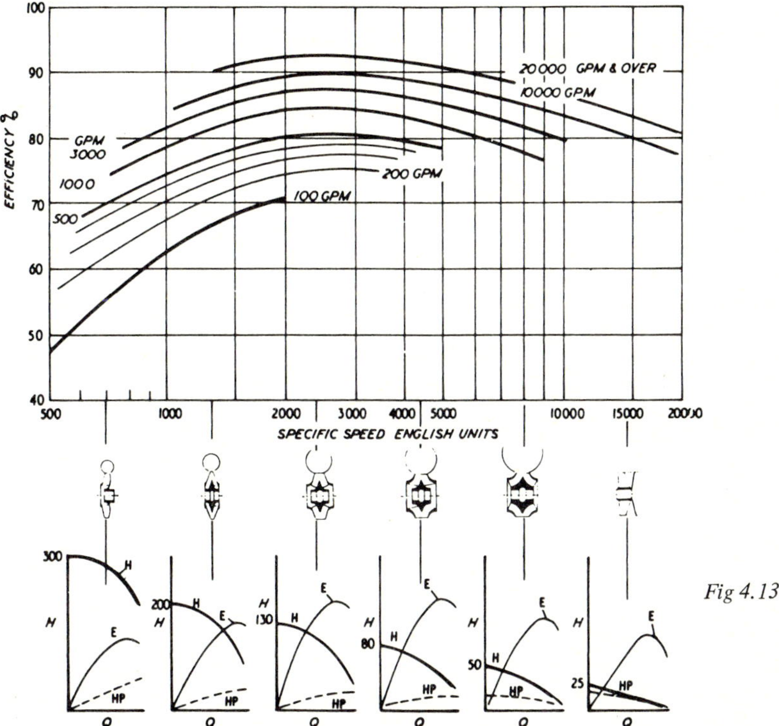

Fig 4.13

This is a generalized illustration only, based on optimum impeller form for the specific speed covered. In practice impeller forms may be modified (*eg* for solids handling) which can satisfactorily reduce efficiency and modify the H-Q curve as well as shifting the design operating point.

Specific Speed and Cavitation

Virtually any type of pump can be made to cavitate by causing it to be operated under insufficient NPSH, but the limitations implied are most significant in the case of rotodynamic pumps since these involve a difficult to determine dynamic depression head within the pump.

In practice this means that for a given total head the required NPSH increases with increasing specific speed, indicating an increasing tendency towards cavitation and thus a reduction in the static suction lift available. This is consistent with the general pattern of behaviour of rotodynamic pumps in that those with the lowest specific speeds have the highest suction lift capabilities, and vice versa. Equally the minimum NPSH is that value at which the cavitation number reaches a critical value and cavitation is imminent. Thus the required NPSH for a pump must be at least equal to, and preferably above, NPSH minimum.

Cavitation

The subject of cavitation is extremely complex and largely indeterminate by fundamental analysis. The onset of cavitation will occur well before the cut-off point, the actual cut-off point being determined by the shape of the impeller and detail design of the pump as well as the specific speeds. At the other extreme, in the case of axial-flow pumps, delivery may be sustained beyond the onset of fully developed cavitation.

Provided the pump is operated within its recommended working range below the cut-off point, cavitation should not be a problem with a good design of pump unless the available NPSH is too low, or if flow conditions depart markedly from the normal. In practice, cavitation may be noticed by noise and vibration, and by mechanical erosion-corrosion on the impeller blades. This is particularly likely to occur where pumps are operated in a region of controlled or limited cavitation, which condition may be acceptable under certain circumstances. In this case specific advantages can accrue from the choice of materials resistant to cavitation and close attention to production technique, notably in the matter of the smoothness of the interior surfaces and surface hardness. In

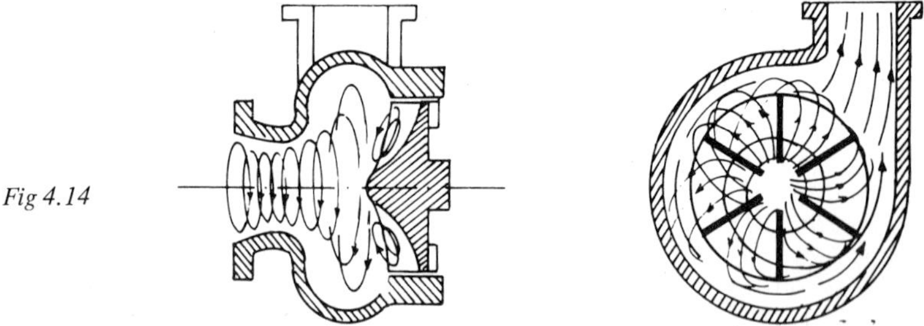

Fig 4.14

the former respect an approximate listing of materials with diminishing resistance to cavitation in acid-resisting steels, stainless steels, carbon steels, aluminium, bronze, cast iron, lead. Pump design factors which increase resistance to cavitation include the shaping of the suction branch and suction chamber to give a smooth fluid entry with gradually increasing velocity, and an induction geometry which imparts a certain amount of pre-whirl to the fluid (in the same direction as the impeller rotation). This is also a feature of some designs of pumps for handling liquid-solid mixture – **Fig 4.14**.

Cavitation Factor

The cavitation factor (σ) is derived as the ratio of the dynamic depression head to the total head of the pump. The former is a quantity specific to the design of the pump and its specific speed. The cavitation factor is thus directly related to the specific speed and also to the efficiency of the pump, but can only be determined empirically or semi-empirically, although a basic formula can be derived, *viz:*

$$\sigma = K.N_s^{\ 1.33}$$

where

N = specific speed

K = is a constant depending on the type of pump and the units in which the c

where

N_s = specific speed

K is a constant depending on the type of pump and the units in which specific speed is calculated.

Strictly speaking, K is not a constant but will vary with the efficiency of the pump at the point of working. It will also differ for a single-inlet pump as compared with a double-inlet pump.

For use with true specific speed, typical values of K are

	English units	Metric units
specific speed	$\frac{0.0174\sqrt{\text{gal/min-}}}{(\text{ft})^{0.75}}$ x rev/min	$\frac{0.1155\sqrt{\text{lit/min}}}{(\text{m})^{0.75}}$ x rev/min
K =	$\frac{75}{\eta^3 \times 10^5}$	$\frac{13.4}{\eta^3 \times 10^5}$

where

η = hydraulic efficiency of pump

To allow for the fact that the value of σ chosen may be far from exact the formula figure would best be factored by 0.8 to 0.9 (depending on the estimated correctness of σ).

Minimum Safe Flow

In the case of a centrifugal pump, power absorbed falls with decreasing delivery and is usually less than one half the design point power at zero flow. However, at lower deliveries a considerable rise in liquid temperature is likely which could lead to flashing, resulting in the pump becoming vapour bound. To avoid this possibility a *minimum safe flow* may be specified, determined by the allowable temperature rise. This, in turn, will depend on the suction conditions notably the NPSH and the temperature and vapour pressure of the liquid. This is normally determined empirically.

Apart from this limitation, centrifugal pumps are capable of being operated continuously, if necessary, down to zero flow. With low specific speeds, however, the H-Q curve may be very flat in this region which may lead to operating difficulties.

In addition to minimum flow requirements determined by temperature, volute type centrifugal pumps tend to develop very large unbalance forces at low flow rates due to the irregular flow pattern established. In extreme cases this can lead to flexing of the shaft and rubbing contact developing. This minimum speed rating may be higher or lower than that established by fluid temperature considerations. Volute pumps fitted with diffuser vanes are largely free from such troubles as the vanes provide smoothing of the flow even at low discharges.

Effect of Speed (rev/min)

Within limits, the effect of running a centrifugal pump at some other speed than its design speed is to modify the head, discharge and power absorbed in proportion to the speed ratio, *viz:*

$$\frac{Q_1}{Q_2} = \frac{\text{rev/min}_1}{\text{rev/min}_2}$$

$$\frac{H_1}{H_2} = \frac{(\text{rev/min}_1)^2}{(\text{rev/min}_2)^2}$$

$$\frac{BHP_1}{BHP_2} = \frac{(\text{rev/min}_1)^3}{(\text{rev/min}_2)^3}$$

The above relationships assume that the efficiency remains unchanged and may be taken as holding approximately true provided the new speed does not differ by more than 25% from the design speed. More exact data can only be determined empirically.

In terms of H-Q characteristics, the effect of a change of speed is to displace the curve upwards or downwards, whilst retaining the same general shape, each speed thus having its particular H-Q curve – **Fig 4.15.** If efficiencies are also plotted on the same graph these will take the form of oakleaf curves or *isoefficiency curves,* providing a complete set of data for evaluating the performance of that particular pump over a range of speeds covered by the analysis. A further curve can be plotted of parabolic shape with its apex at the origin and cutting the other curves at the points representing optimum operating conditions at the particular speeds concerned, *ie* maximum efficiencies attainable at each speed. This curve can also be plotted separately against speed of discharge, if desired – **Fig 4.16** – to indicate maximum efficiencies attainable at any particular speed or discharge.

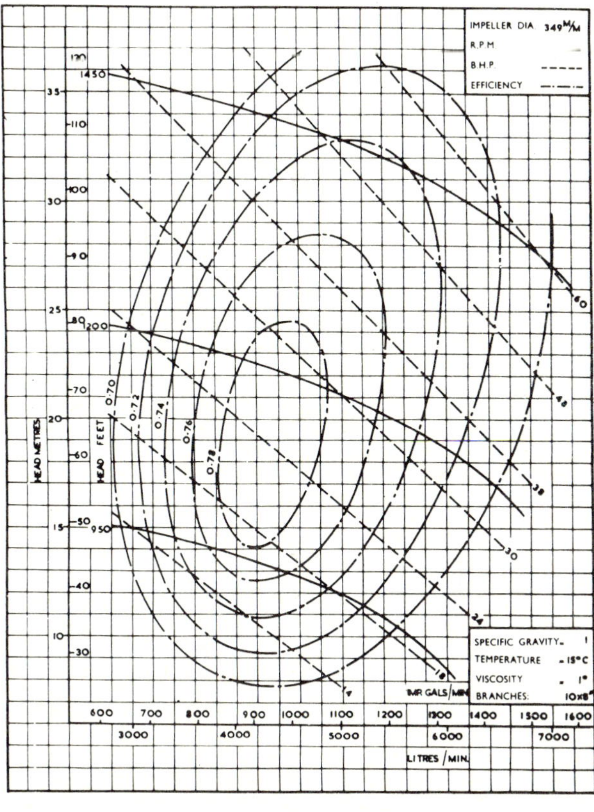

Fig 4.15 Typical speed, isoefficiency and BHP curves.

Working Range of a Centrifugal Pump

Although a centrifugal pump achieves maximum efficiency only at a specific working point the potential working range is obviously much higher, *ie* it can be extended in either direction along the H-Q curve. The potential range can also be extended by increasing or decreasing the operating speed, displacing the H-Q curve upwards or downwards, respectively.

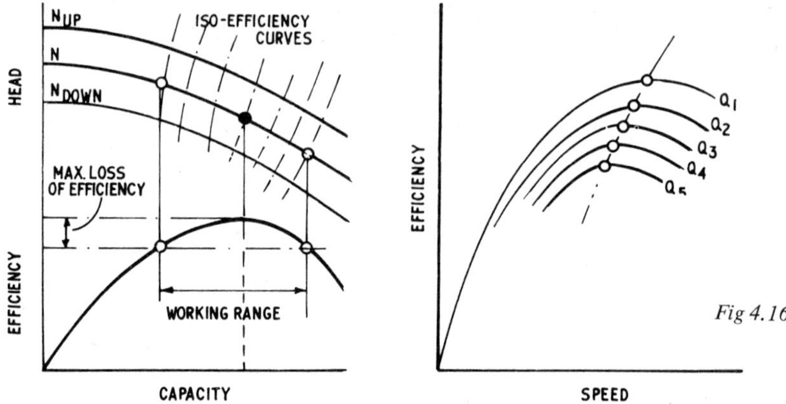

Fig 4.16

A practical field of application can then be set by:

(i) a limit to the minimum efficiency desirable;

(ii) upper and lower limits to the speed change.

A limit to the efficiency will establish extreme points for working along the H-Q curve for the design speed. If additional H-Q curves are then plotted for the two speed limits an envelope of working range or field of application can then be completed by the isoefficiency curves through the limiting points on the design speed curve – see **Fig 4.17.** The pump can then be operated at any point within this envelope with an efficiency not less than the minimum.

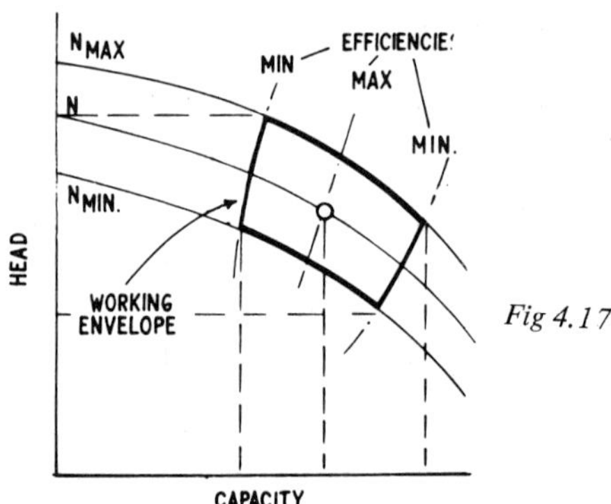

Fig 4.17

A similar type of envelope of working range can be established for a pump designed for operating at a constant speed by substituting H-Q curves for different impeller diameters.

Some points of significance:

(i) The operating range can extend, theoretically at least, over the full length of the H-Q curve, up to the cut-off point if efficiency of operation is not important. However, with pumps having unstable H-Q characteristics the practical operating range would never be considered to extend to lower capacities beyond the peak of the H-Q curve.

(ii) The flatter the efficiency curve the wider this length of working along the H-Q curve will be for the same limit for drop in efficiency. Where an extended working range is required it is thus desirable to select a pump with an impeller design giving a flat efficiency curve. Nozzle-type impellers are usually superior in this respect, but hydraulic efficiency can also be maintained with inducers.

Impeller Diameter

It is common practice to fit different diameters of impeller in a single casing thus giving a range of characteristics to a particular model. The effect is similar to speed variation in that each impeller diameter generates its own H-Q curve – **Fig 4.18.**

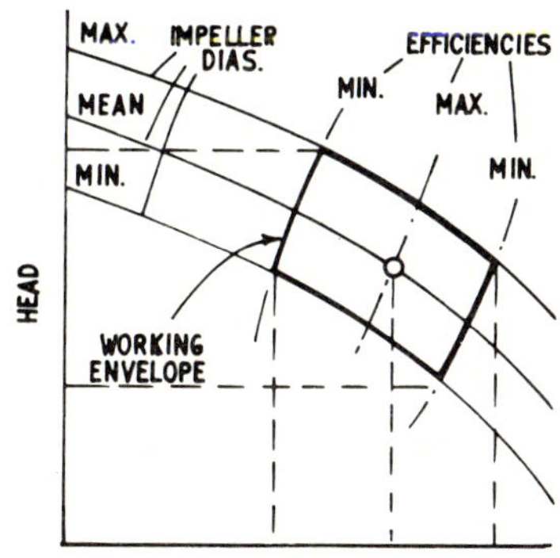

Fig 4.18

Again isoefficiency curves can be plotted on the same graph, but rather than include these as a standard feature they are normally employed only to establish an 'envelope' enclosing a range of practical application of that particular model. The other useful information which can also be included within the envelope is the horsepower absorbed, normally presented as diagonal 'horsepower lines'. Given such a rating chart, the operating range, characteristics (*ie* H-Q) for any particular impeller diameter, and power requirements, can be read.

Composite Rating Charts

Where a line or series of centrifugal pumps is concerned the different models are designed to cover a wide operating range. The working range of each individual model can be designated by an envelope, as above and the performance characteristics of the complete range presented as a series of such envelopes, as in **Fig 4.19.**

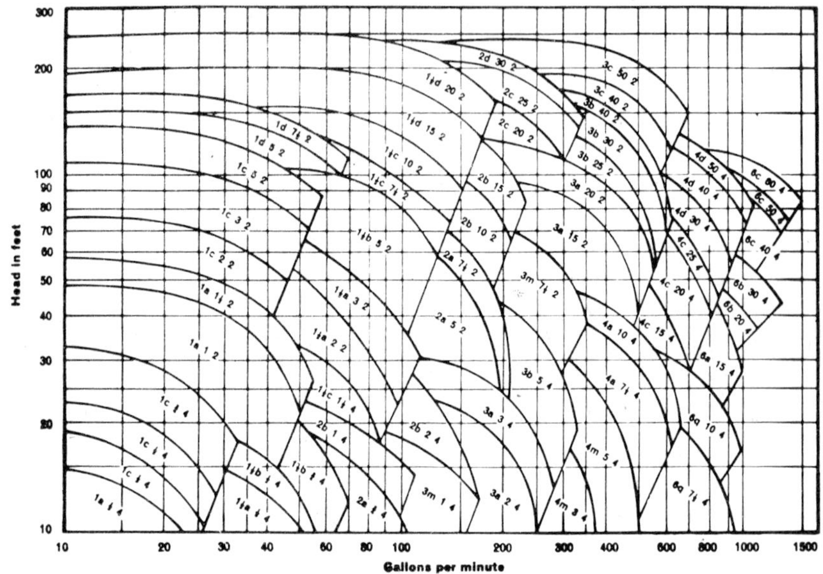

Fig 4.19 Typical composite rating chart.
(Holden and Brooke 'J' range pumps)

Such a composite rating chart defines the optimum working range of each individual model in the series and thus establishes the choice of pump size for any required operating point. Further information can then be extracted, as necessary, from within the envelope of the chosen pump.

The shape of the individual envelopes, and the form of the complete rating chart will, of course, be specific to an individual series of pumps. Thus whilst the overall picture may be similar, a composite rating chart for one manufacturer's series cannot be taken as representative of those of other pumps of similar size.

Rating Tables

Rating Tables present similar data in tabular form, although the picture available is far less complete. The information given is usually restricted to practical capacities with power requirements for different heads, designating, in fact, the H-Q range for efficient working. Some manufacturers give only rating tables for their individual pumps or series of pumps; others give rating charts. In either case, sufficient information is available to select a suitable size of pump for a given working point.

Variable Demand

A centrifugal pump is normally selected on the basis of a fixed working point, *ie* of a size where the required pressure and head corresponds to the design working point of the pump, as any departure from the design working point will tend to lower the pump efficiency. Where demand varies, however, some form of capacity control may be essential which, in the case of a constant speed machine, usually takes the form of throttling the discharge. Adaption of a centrifugal pump to a permanent change in demand, however, may often be best met by a change in impeller.

Throttling can only be applied on the discharge side, because if applied to the suction side it is likely to cause cavitation. It is a generally effective method of control and not unduly wasteful of power since the power input decreases with decreasing discharge. There is, however, a drop in total head, due to the hydraulic resistance of the throttling valve.

In the case of mixed-flow pumps with high specific speeds, and axial-flow pumps, regulation by throttling demands an increase in power and by-pass regulation is preferred. In this case a proportion of the delivery is tapped off through a by-pass and returned to the suction side of the pump.

Various other methods are available for adjusting a rotodynamic pump to meet a change in demand, notably –

(i) variable speed control (particularly applicable where the driver is of a type suitable for speed regulation, or it is economic to use a frequency converter).

(ii) the use of adjustable inlet guide vanes (normally only effective on pumps with higher specific speeds).

(iii) the use of variable pitch impellers (axial-flow pumps only).

(iv) using more than one pump to meet maximum demand and shutting down one or more as demand drops.

(v) the use of a storage tank and intermittent operation of the pump to maintain a suitable tank level.

NPSH

The NPSH required by a centrifugal pump is usually determined by measurement of its water performance and thus refers to the head of water. This NPSH figure will be the same for all fluids, *ie* if the pump is to be used to handle a fluid of different specific gravity exactly the same figure for NPSH applies, except that this can now be taken as referring to feet of the liquid concerned. In other words, NPSH required figures should not be 'corrected' for fluids of different specific gravity, except when the fluid differs appreciably from the standard fluid (water) in thermodynamic properties. In normal practice this is only likely to apply in the case of light hydrocarbon fluids and other volatile fluids which have a vapour pressure at pumping temperature greater than 14.7 lb/in² (1 bar) absolute. In such cases the NPSH required will be reduced, compared with the water figure, typical corrections for light hydrocarbons being given in **Table II.**

TABLE II – NPSH REQUIRED AS PERCENTAGE OF WATER FIGURE

Vapour Pressure * (absolute) lb/in²	bar	Specific Gravity of Fluid* 1.0	0.9	0.8	0.7	0.6	0.5
14.7	1.04	60	73	88	100	–	–
50	3.9	60	68	76	85	94	100
100	7.4	60	65	71	78	83	90
200	14	60	63	67	72	76	80
300	21	60	61	64	67	70	73
500	35	60	60	62	64	66	68

*At pumping temperature

Classification of Centrifugal Pumps

In addition to being a separate class of their own, centrifugal pumps may be further sub-divided or classified in a number of ways. Whilst the majority of these may be intended to simplify selection of a suitable pump for a particular duty, the information given by arbitrary classification is very incomplete. Nevertheless such classifications are widely quoted, and used by manufacturers.

A number of classifications based on pump characteristics have already been mentioned – *eg* high, medium and low head types applied to centrifugal pumps. Other methods of classification may be based on general recommendations, application or duty, or design and geometry.

Illustration shows typical low NPSH extraction pump. (Girdlestone Pumps)

TABLE III – CENTRIFUGAL PUMPS FOR SPECIAL APPLICATIONS*

Service or Application	Duty	Impeller	Special Design Features	Remarks
Raw sewage	General	Single-blade	Circular casing	Unchokeable impellers No diffusers
	High capacity	2-channel	Volute casing	Comminuting device or macerator may precede impeller
Screened sewage	General	2-channel 3-channel	Flushed gland Flushed gland	
	High capacity	Mixed flow	Flushed gland	
Mud	General	2-blade	Flushed gland	Lined pump or replaceable wear parts
Fibrous materials	General	2-blade	Comminuting device or macerator may precede impeller	
Fermenting liquors	General	Open 2-blade Open 3-blade Ejector type	Suction chamber mounted above impeller	Provision required to 'Gas Off' suction pipe
Solids	General	1-3 blade single or 2-channel	Flushed glands; wear protection	Design matched to size and shape of solids
	Soft solids	Special open	Flow sections must be generous	*eg* Fish, fruit *etc*
	Sugar beet	Square or 2-blade Special	Flow sections not less than inlet and outlet branches	
	Abrasive	1-3 blade single or 2-channel	Lined or armoured casing; replaceable wear parts	Clearances according to size and shape of solids
Stuff (Pulp, *etc*)	Less than 3% (dry content)	Standard open type		Performance derated compared with water
	3-5% (dry content)	Open		
	Over 5%	Ejector type		
Chemical	General	Standard	Materials must be compatible	
	Abrasive	1-3 blade single or 2-channel	Materials must be compatible	
	Corrosive	Standard	Materials must be compatible	Glandless pumps are often preferred

*See also Chapter 9.

Classification by Duty (see also **Table III**)

This can be a specific guide for certain applications where the requirements are rigid and clearly defined since the pump is usually specially designed or modified to meet the operating conditions involved. Other application categories admit a much wider range, however, and are thus far less specific as a selection guide. Thus the fact that a particular machine may be classified as a chemical pump, or a process pump, for example, does not necessarily preclude the possibility that it may have to be modified as regards materials of construction or gland design, or both, to handle a particular fluid successfully. Also the primary fact remains that it is the performance of the pump that ultimately counts and this which finally decides the suitability or otherwise of a particular pump type. There may also be other practical factors involved, such as ease of dismantling for inspection or cleaning, overall bulk and weight, and so on.

Provided material compatibility is achieved in the wetted parts of the pump, centrifugal pumps are capable of handling virtually all clean liquids of low viscosity at ambient or moderate temperatures. Where no specialised attention is given to corrosion resistance, the centrifugal pump is a true general purpose type, and most efficient in the larger sizes or where high discharge rates are required. Essentially the same layout built in corrosion resistant materials would be classified as a chemical pump; and one with more robust construction to resist erosion-corrosion, a process pump.

Impeller Forms

As previously noted the form of the impeller is largely dictated by the specific speed of the pump. A variety of different shapes and arrangements of the blades can be produced within a particular form, a main classification being whether the impeller is shrouded (closed) or unshrouded (open). Closed impellers are common to centrifugal pumps (and some mixed-flow pumps), with shrouds rigidly attached on both sides to enclose the liquid passage. However, for the handling of fibrous materials in suspension in the liquid the impeller may be semi-open (*ie* with a shroud on one side only) or fully open, or a skeleton impeller rotating between stationary discs – see also **Fig 4.20**. The use of semi-open or open impellers together with modifications of blade shape can make the centrifugal pump suitable for handling contaminated rather than clean fluids.

Impeller designs may be further modified by the method which the liquid enters the impeller.

Changing the number of blades, or the form of the blades, will affect the performance characteristics. Increasing the number of blades will tend to flatten the H-Q curve; and vice versa. For a given number of blades, increasing the curvature of the blades will tend to steepen the H-Q curve, and vice versa. Decreasing the width of an impeller will also tend to steepen the H-Q curve, and increasing the width, to flatten it. These effects are general characteristics only and will be partly modified by changes in specific speed which may result. They serve to

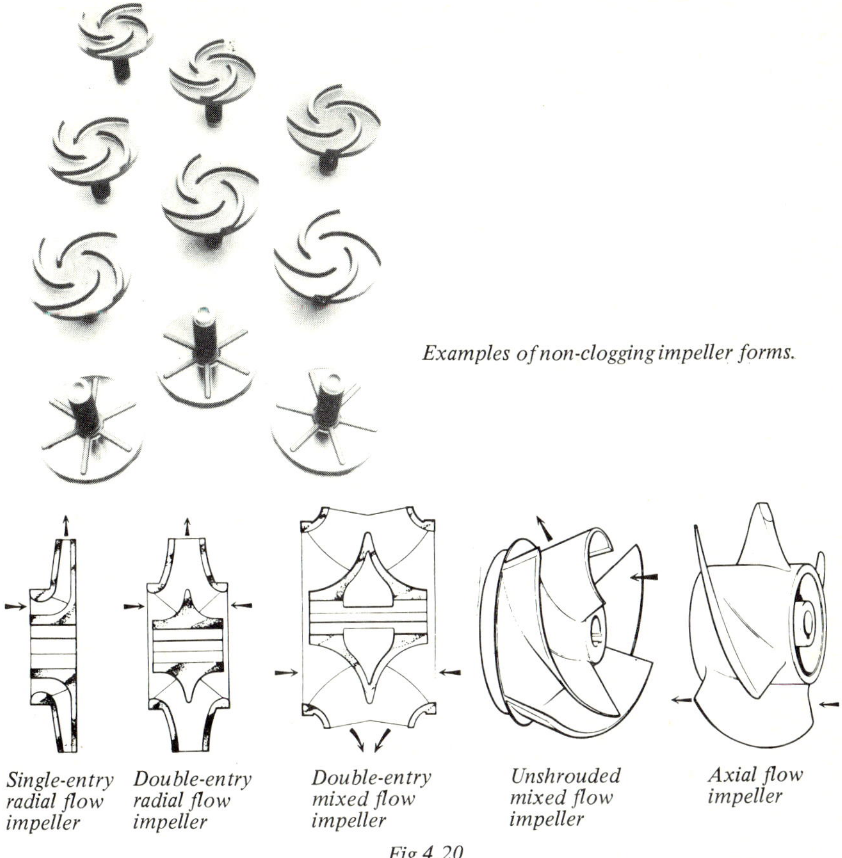

Examples of non-clogging impeller forms.

Fig 4.20

indicate, however, that with the considerable variation possible in blade shape, form and number, the resulting H-Q characteristics are essentially individual to that particular design of pump.

SPECIAL DUTY IMPELLERS

The ability of a centrifugal pump to handle other than clean, uncontaminated fluids is largely governed by the impeller design. The variety of different forms available for other services are most conveniently described under the specific, or intended, application.

Unchokeable Pumps

These are generally intended to handle the widest variety of liquid-solid mixtures although the impeller form is often specifically matched for handling a particular type or size of solid contaminant. Impellers are almost invariably of open type with a limited number of blades and side entry is usually preferred. The general purpose 'unchokeable' pump may be designed to handle a variety of solid contaminants up to a specified size, which can be as large as the suction branch diameter (provided the same clearance through the rotor and discharge branch is maintained). In practice, however, the limit to the size of solids may be set by the water velocity attained and its consequent carrying power.

For handling fibrous solids special designs of impeller are often employed which may also include auxiliary cutting blades, or be preceded by a separate communiting device. Some typical impeller forms are shown in **Fig 4.21**, but these are intended as general illustrations only as there are numerous individual and patent designs (especially for the handling of fibrous solids).

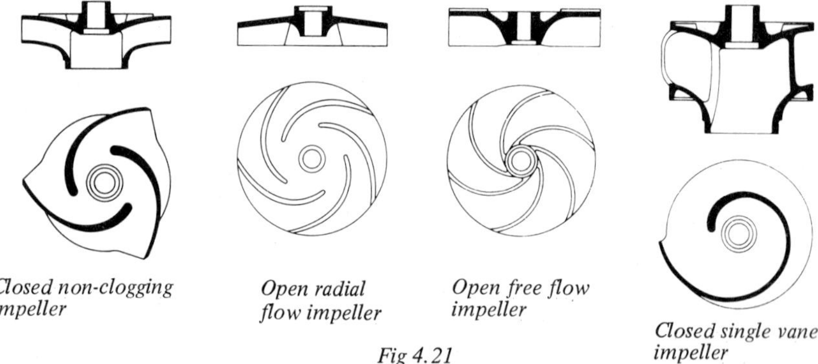

Fig 4.21

The rest of the pump itself may also need considerable modification in detail design, notably both to resist abrasion and wear caused by the presence of solids, and possible unevenness of flow. The former can be met with armouring, lining and/or the use of replaceable wearing parts. Provision is also usually made via detachable cover plates to gain ready access to the impeller for cleaning, if necessary.

Ejector-type impellers are a special type which show high efficiency when handling entrained gases, pulp and also paper stock and similar dry solids in suspension. Typically these have from two to five non-overlapping blades with thick inlet edges, the impeller passages formed producing a low pressure region on the back face of the blade rendering the impeller relatively insensitive to the presence of entrained air in the pumped fluid.

In general, however, special impeller forms for handling mixtures of solids and liquids inevitably show much lower efficiencies than conventional impellers (although the latter would not be suited to the duties involved). Where conventional impellers could be employed, performance is derated compared with that of the same pump and impeller handling water.

Special impellers have also been developed for the handling of fragile solids, such as foodstuffs, fish, fruit, *etc,* which must be passed through the pump without damage. Flow passages are generously proportioned and the flow so established as to carry the solids through the pump without actually contacting the impeller.

Entry Geometry

In the simplest form of centrifugal pump the suction branch is mounted on the axis of the pump and opens into the central eye of the impeller. The liquid is then flung out under centrifugal force into a volute casing terminating in a recuperator section, to the end of which is attached the delivery flange. The impeller itself is overhung and the single entry flow imparts an axial thrust on the impeller, which can be partially relieved, if necessary, by holes in the back shroud. The balance of the axial thrust is then taken by the shaft bearings.

With a double entry pump the inflow is divided and taken to each side of a double sided impeller, thus balancing the axial thrusts. Pumps with double-entry are usually of larger size, with higher specific speeds. Suction and delivery branches are commonly situated on one half of a split casing, the other half being removable for inspection or for dismantling the pump without the necessity of first disconnecting the suction and delivery pipes.

As a general characteristic the suction branch diameter size is a little larger than the delivery branch on centrifugal pumps. This, basically, is to encourage the use of a generous size of suction pipeline to minimise suction losses; but in practice this does not necessarily determine the corresponding pipe sizes. Delivery and suction branch sizes are normally standardised (to mate with standard pipe flanges, or with standard pipe threads in the case of small pumps). Optimum pipe sizes may be larger, when reducing pieces will have to be employed to connect piping to the pump flanges.

Classification by Overall Geometry

An immediate classification is given by the number of stages (or separate impellers mounted on a common shaft) contained within the casing. Thus a single-stage pump implies a single impeller (although in some cases it may be double-sided) and a single stage of pressurisation of the liquid passing through the pump. Pumps with more than one stage are generally referred to as multistage machines, with the number of stages appended where applicable. Thus in the case of unit construction often employed for multi-stage pumps, any number of stages may be added, within practical limits, by bolting on additional sections.

A further geometric classification depends on whether the pump is designed for horizontal or vertical mounting. For general purpose applications the choice is somewhat arbitrary. Horizontal pumps are the more common and have advantages from the point of view of ease of maintenance, resistance to corrosion and abrasion and ease of mounting with minimum headroom needed. Vertical pumps have the advantage of occupying minimum floor space, ease of priming and reduction in required NPSH and general flexibility to adapt to a particular duty. With certain types of pumps the application may dictate mounting – *eg* submersible and immersible pumps are invariably of vertical type.

In manufacturers catalogues, pumps are commonly described by geometric characteristics, *eg* horizontal, single-stage, single-entry centrifugal pump. The general form, and to some extent the limits of performance and general application can be derived from this. Classification and identification on this basis is summarised in **Table IV**.

TABLE IV – CLASSIFICATION BY GEOMETRY

Basic Type	Stage(s)	Layout	Entry	Casing
Centrifugal	Single-stage	Horizontal	Single	Volute-split or end pull-out
	Single-stage	Horizontal	Double	Volute-split or end pull-out
	Multi-stage	Horizontal	Axial (single or double)	Ring type for low to medium heads. Barrel type for high pressures
	Multi-stage (balanced)	Horizontal	Back-to-back	Volute (usually)
	Single-stage	Vertical	Single	Volute-split casing
	Single-stage	Vertical	Double	Volute-split casing
	Multi-stage	Vertical	Single	Volute
Mixed flow (helical)	Single	Horizontal	Single	Volute-split casing or end pull-out
	Single	Horizontal	Double	Volute-split casing
	Single	Vertical	Single	Volute-split casing
Mixed flow (diagonal)	Single	Horizontal	Single	Concentric
	Single	Vertical	Single	Concentric
	Multi	Usually vertical	Single	Concentric
Axial flow	Single	Horizontal	Single	Concentric
		Vertical	Single	Concentric
	Multi	Usually vertical	Single	Concentric

Classification by geometry also extends to the manner in which the casing can be opened for disassembly, *eg* split-casings or end pull-out. In the case of multi-stage pumps the method of assembly of the individual stages is usually specified – *eg* ring-type or barrel-type (*eg* **Fig 4.22**). The latter type are normally used for high pressure designs, the casing consisting of an inner casing fitting within an outer barrel. Discharge pressure acting on the inner casing provides a sealing force to hold the casing halves together, whilst the construction also minimises the number of connections and sealing points required.

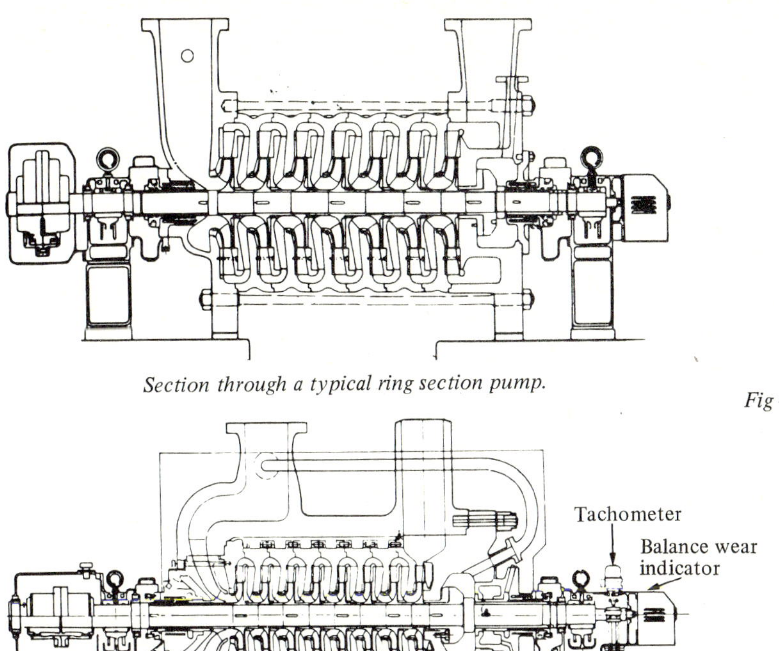

Section through a typical ring section pump.

Fig 4.22

Section through a typical barrel casing pump.

Self Priming Characteristics

With one exception, rotodynamic pumps have an inherent inability to pump out air from the pump casing and suction line and will therefore not self prime on their own. The exception is the side-channel (regenerative) pump with its turbine wheel impeller form since suction and delivery lines are invariably mounted on the top of the casing and thus a residual charge of liquid always remains in the side channel when the pump is stopped. This is sufficient to act as a prime when the pump is restarted, although this type of pump will still require priming with liquid when starting up for the first time.

To overcome the lack of self-priming ability with other types of centrifugal pumps they must either be operated with flooded suction or be associated with a suitable priming device to make them self-priming. The four main types of devices will be mentioned briefly, although there are also other possibilities.

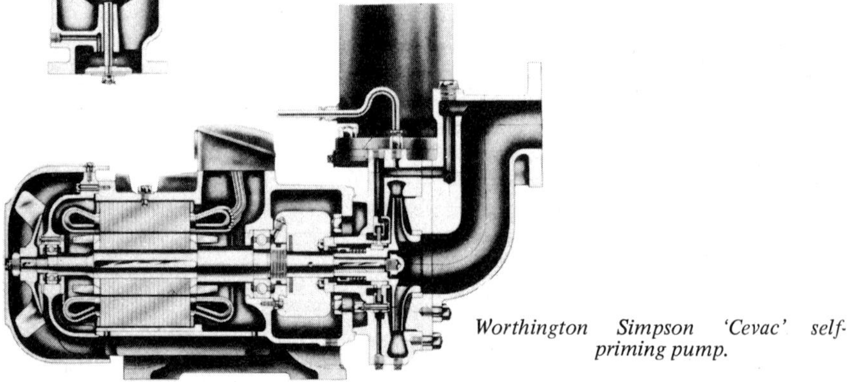

Worthington Simpson 'Cevac' self-priming pump.

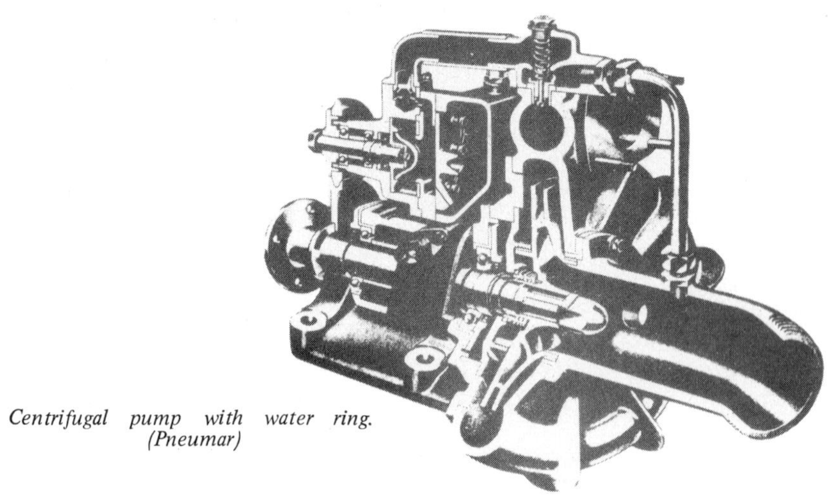

Centrifugal pump with water ring. (Pneumar)

(i) **Priming Nozzles** – Provision is made in the design of the pump to trap a residual charge of fluid within the pump when the pump is shut down. This normally involves the use of a modified casing shape with the suction branch at the top and a non-return valve to prevent siphoning back through the suction line, or a separate residual liquid chamber attached to the pump.

On starting up again a continuous flow of this residual liquid is generated through a priming nozzle directed into the eye of the impeller. The flow from the nozzle is broken up into a spray, sweeping through the impeller and entraining air. Entrained air escapes through the delivery pipe and the liquid is recirculated until all the air has been discharged and normal fluid flow is established up the suction pipe, through the impeller and out through the discharge. Flow through the priming circuit is then stopped by manual, or automatic operation of a shut-off valve.

Other designs of self-priming pumps may have special volute designs or guide vanes to promote a 'nozzle' effect to generate a liquid-air mixture in the impeller passages when the pump is started up. Separation of air is then achieved in a nozzle section, air passing into the discharge branch and liquid returning for re-circulation. This continues until all the air has been evacuated from the suction pipe, when full fluid flow is maintained. Depending on the form of the nozzle device which remains in circuit the efficiency of the pump under normal working is then reduced accordingly, although with good design the loss can be very slight.

(ii) **Use of separate air-exhausting pump** – This self-priming system employs a secondary pump combined with the main centrifugal pump. This is commonly a liquid ring pump (which may be in tandem on the same shaft) or a peripheral pump which may be incorporated on one face of the centrifugal pump impeller, when provision must be made to retain a certain residual charge of liquid with the attendant side channels. The purpose of the secondary pump is to evacuate the air within the pump casing and suction line, the centrifugal pump taking over as soon as the casing is filled with liquid, although both pumps will continue to work in parallel.

(iii) **Use of Diffusers** – The principle is similar to that of the priming nozzle, provision being made to retain a residual charge of liquid in the casing when the pump is stopped. On starting up again this liquid is circulated through diffuser passages to impinge on the impeller and be turned into spray. Air is then removed by entrainment, as with the nozzle system. When the pump is running normally the diffusers merely act as a recuperator.

(iv) **Use of ejectors** – Definitive ejector pumps or jet pumps may also be used as auxiliary devices to promote self-priming in centrifugal pumps. A certain level of liquid is retained in the ejector circuit. On starting up, flow is established in the ejector circuit, drawing in air from the suction pipe. Air passes through the pump to be discharged on the delivery side (usually separating out in a reservoir mounted on the pump outlet) whilst circulation continues until the suction pipe is completely evacuated and normal pumping is initiated (completing discharge of air from the reservoir, if fitted). Since energy losses are involved by the flow through injectors, a self-balancing circuit is often employed in such cases, automatically stopping circulation through the ejector circuit once normal pumping is achieved.

Ejectors have the advantage that they can generate quite high suction heads and are generally insensitive to impurities in the liquid being handled, unless

End suction centrifugal pump in all-plastic construction except for shaft and seal.
(ASM Pacer Pumps)

Large double suction centrifugal pump – top valve of casing removed.

these have a clogging action. In the case of solid contaminants a settling tank may be used on the suction side to reduce the possibility of solids entering the ejector circuit.

The above are essentially self-priming systems, *ie* the pump, so fitted, has self-priming characteristics. The same end can be achieved by ancillary priming devices which may normally be associated with the driver rather than the pump itself, or by a separate mechanical, electrical or hydraulic system. A distinction is thus drawn between a self-priming pump (in which the necessary device to promote self-priming is incorporated in the pump itself), and ancillary priming devices, which may be employed to supply prime to a pump lacking self-priming characteristics.

5. Positive Displacement Pumps

PUMPS, other than rotodynamic types, work on the more obvious principle of positive displacement of the fluid being handled, and as a consequence have quite different characteristics to rotodynamic machines. They are known generally as positive displacement pumps and can be sub-divided into two main categories – (i) reciprocating pumps, and (ii) rotary pumps – each of which gives rise to a number of specific sub-types. The above classification is not complete for there are a limited number of non-rotating devices which do not have positive displacement characteristics, and also special types of pumps or pumping devices which provide positive throughput but cannot properly be described as either reciprocating or rotary machines.

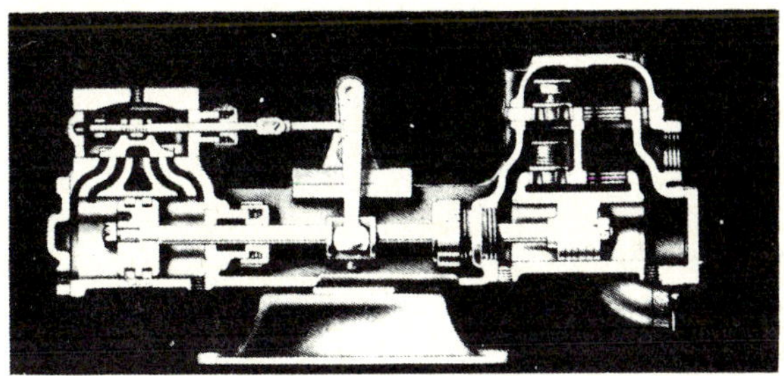

Direct-acting reciprocating pump.
(Worthington Simpson)

RECIPROCATING PUMPS

The basic reciprocating pump comprises a piston or plunger which is reciprocated in a cylinder. The resulting capacity is the product of the bore area and length of stroke and the speed of stroking. The practical capacity is reduced by any internal slippage present, *ie* due to leakage past the piston, or through any by-pass arrangement incorporated in the design. With good design, volumetric efficiency can closely approach 100% after an initial run-in period. Thus the full theoretical value is commonly used in estimating the capacity of a reciprocating pump at any particular speed, although in practice a figure of 95–98% of the theoretical capacity is more realistic for pumps in good condition. Power losses are due mainly to friction and thus reciprocating pumps are inherently capable of achieving high overall efficiencies. Overall efficiences may approach 90% in larger pumps, but tend generally to decrease in line with decreasing pump size, although this can be offset by more precise construction.

The terms piston, plunger and ram are synonymous as describing the inner reciprocating element, although they are used separately to describe a particular type of pump. Thus 'piston' is the description normally used in the case of larger pumps with larger bore sizes and high capacities; whilst 'ram' may be used for similar size ranges specifically designed to develop high heads or delivery pressures. 'Plunger' is used in the case of small bore high pressure pumps, and specifically so where the element differs physically from a true piston. Thus a piston is mounted on a piston rod, whereas a true plunger is a rigid entity reciprocated by cam motion, or similar. Thus, in the description of a particular pump, the word 'ram' commonly implies that the pump is of high capacity, high pressure type; and the word 'plunger' a low capacity, high pressure type. Ram pump is also the description applied to slow moving heavy-duty reciprocating pumps (sometimes hydraulically operated) designed to handle semi-solid and near-solid products. These classifications are arbitrary, however, and not universal.

Reciprocating pumps may also be classified as being either direct-acting or power driven. With a direct-acting pump the driving power is derived from steam and a steam piston and pump piston are mounted on a common rod. With a power driven pump the power is derived from an external driver and (usually) a crank mechanism to produce the reciprocating motion.

Discharge Pulsations

Although the capacity is the same for each type (for the same size of pump) the delivery characteristics are quite different. Thus with a direct-acting pump the delivery is in the form of a steady flow with abrupt steps at each end of the

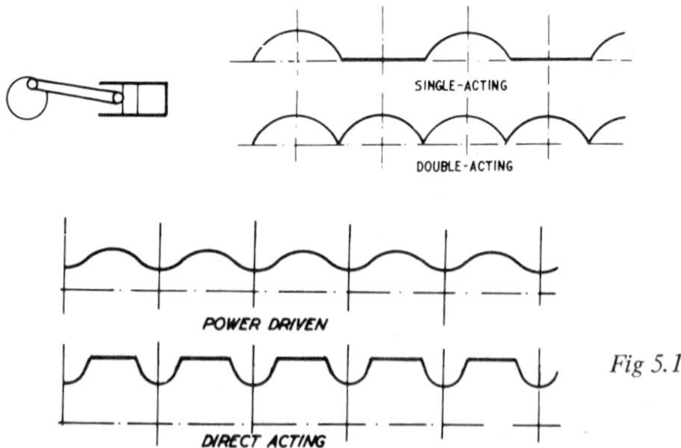

Fig 5.1

stroke where the piston stops and reverses its direction of motion. With a crank-operated power driven pump the discharge is in the form of sine waves – **Fig 5.1**. Thus whilst both types produce pulsating outputs the nature of the pulsations is quite different; and the effect of pressure variations is far more significant in the case of power driven pumps, because the latter inherently tend to produce pressure surges.

It is, however, readily possible to provide smoothing of the pressure fluctuations by incorporating an air chamber or cushion chamber on the delivery side – *eg* **Fig 5.2.** This is simply a closed air vessel connecting to the delivery side of the pump and charged with compressed air (either from an air compressor in the case of large pumps, or from an automatic air valve on the pump cylinder). The effect is to smooth out the delivery flow, as shown in **Fig 5.3**, due to the air 'cushion' established in the circuit. Pumps so fitted may be referred to as cushioned types. In some cases a cushion chamber may be fitted on the suction side, this usually being equally effective. Alternatively a separate *pulsation damper* or *surge suppressor* may be incorporated in the system (usually close to the pump).

Smoothing can also be effected by employing more than one cylinder and overlapping their respective strokes. Thus **Fig 5.4** shows typical discharge characteristics for Duplex (two-cylinder) direct-acting pumps with the discharge of one

Fig 5.2: Reciprocating pump with air cushion chamber.

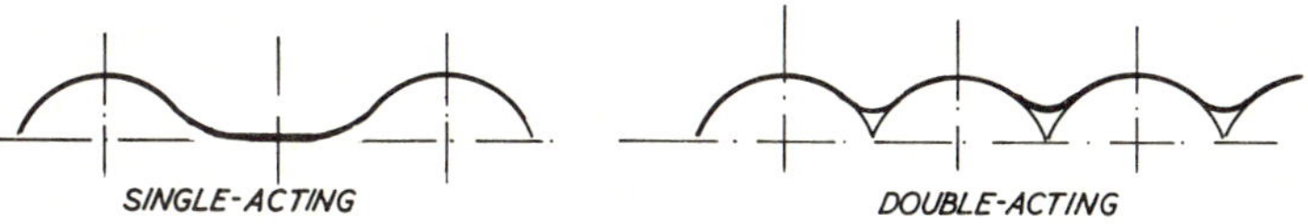

Fig 5.3

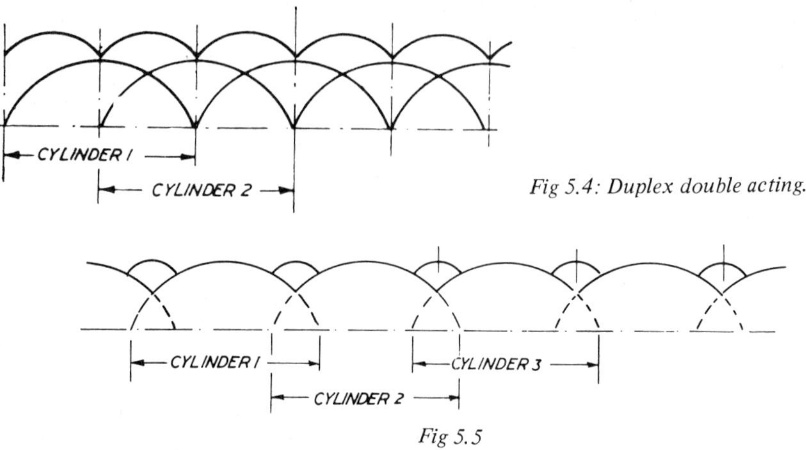

Fig 5.4: Duplex double acting.

Fig 5.5

cylinder displaced half a stroke from the other. **Fig 5.5** shows corresponding curves for Duplex and Triplex (three-cylinder) double-acting power driven pumps. In all cases, of course, further smoothing can be provided by air cushioning so that with multiple-cylinder construction pressure pulsations can usually be reduced to insignificant levels. A further advantage in such cases is that the input torque required is much more uniform.

The actual degree of pulsation present in the delivery will also depend on the detail design of the pump, notably the relationship between connecting rod length and stroke, or crank angle, and also the type and response of the valves.

H-Q Characteristics

Theoretically, the H-Q curves for reciprocating pumps are represented by vertical lines, each line corresponding to a particular speed. In practice there will usually be some departure from the vertical since slippage will tend to increase with increasing delivery pressure – **Fig 5.6.** There is no theoretical limit to the pressure which can be developed, other than by the input power available and the strength of the pump components to withstand that internal pressure or working load. In practice, this means that the reciprocating pump is capable of building up dangerously high pressures if operated against a heavily throttled discharge and so would need protecting by a pressure relief valve (safety valve) if worked under such conditions. The direct-acting pump differs from the power driven pump, however, in that there will be a limit to the pressure it can develop set by the steam pressure available. Thus the pump will only continue to work as long as there is a pressure difference available for driving the pump, and if the delivery pressure goes on increasing the pump will eventually stall.

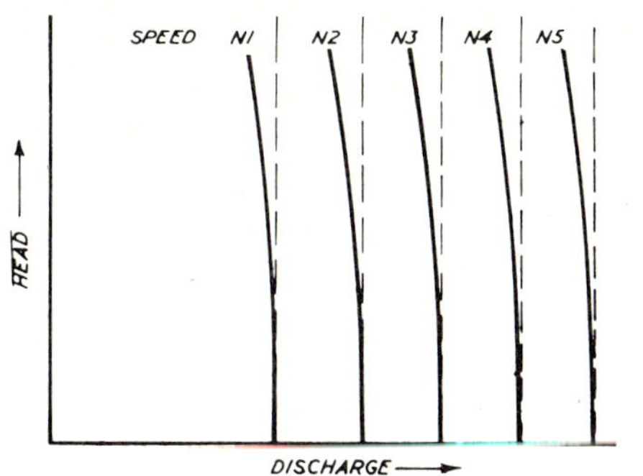

Fig 5.6

Discharge Characteristics

Discharge is directly proportional to speed and will go on increasing with speed. The limit is set by mechanical considerations, or the speed at which there is insufficient time for the suction side of the pump to fill completely with a charge of fluid. The most significant parameter in the latter case is the fluid viscosity and it is commonly necessary to reduce the speed of a reciprocating pump when handling liquids of higher viscosities – see **Chapter 8.**

Speed limits are also set by the size of pump, and in particular acceptable limits for piston speed and the design of valves suitable for high speed operation. Working limits generally fall within the envelope defined by **Fig 5.7** with higher speed operation being restricted to shorter strokes, so that the higher speed reciprocating pump is not necessarily a higher capacity unit.

In practice, the size of any particular reciprocating pump is largely nominal since the capacity it can achieve will be determined by the speed at which it is operated, with an upper limit to speed set by the practical factors mentioned above. Thus a capacity rating is appropriate only to a stated speed. The limit

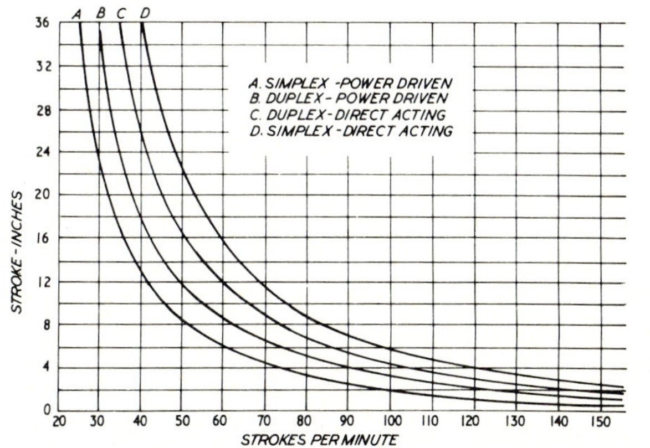

Fig 5.7

for the practical working pressure is more clearly established by the design. Thus the working range or duty range of a particular reciprocating pump can be defined by an envelope bounded by the maximum design pressure and the design limiting speed, such as shown by **Fig 5.8.** The working point can then be anywhere within this envelope without greatly affecting overall efficiency, except at low heads.

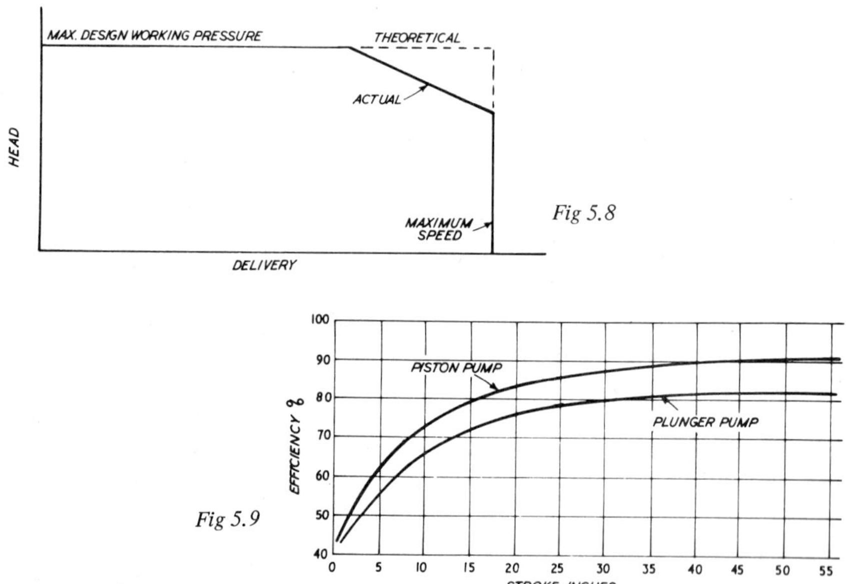

Fig 5.8

Fig 5.9

Efficiency

Efficiency is not readily determinable, except by direct measurement. Hydraulic horsepower output is directly proportional to delivery x head and thus for a given speed (delivery) input power required is directly proportional to head. Efficiency, however, at first rises rapidly with increasing head and thereafter remains substantially constant. There is thus no optimum working point, as is the case with rotodynamic pumps, and so reciprocating pumps can be generally worked over a range of head with little change in efficiency. Efficiency figures should also not vary greatly with changes in delivery (different speeds), but can vary markedly with different types of pumps – see **Fig 5.9.**

Variable Delivery

Where variable capacity is required from a reciprocating pump this can be achieved by (i) altering the speed, or (ii) altering the stroke. The former is not practical where a constant speed driver is employed, unless the system warrants the use of a variable-speed transmission link or frequency converter. Speed will, however, be controllable by steam pressure in the case of a direct-acting pump. Variable stroke devices are a practical alternative fitted to the pump itself and widely used where it is desirable, or necessary, to vary the delivery of a constant speed reciprocating pump. Other methods which may be used include variation of valve timing and the incorporation of by-pass devices. Discharge regulation by throttling is not practical with reciprocating pumps or any type of positive displacement pump.

Seals and Packings

All reciprocating pumps require gland seals or packings and the piston rod gland may present limitations for the applications of such pumps. Thus all rod seals normally rely on a small but positive leakage to maintain satisfactory lubrication and low friction, the question of what is an acceptable level of leakage becoming more difficult as the output pressure is increased. This may call for an extremely long or complicated gland. Also if the fluid being handled is not a satisfactory lubricant the gland may have to be lubricated from a separate source with the risk of contaminating the product being pumped (since gland leakage will occur both inwards and outwards). Equally, to prevent leakage of a toxic, corrosive or hazardous fluid, the gland may be flushed with an inert fluid when the only external leakage will be of inert fluid, although the product itself will also be contaminated with the same fluid.

The difficulty of producing satisfactory gland sealing without resort to complicated or expensive designs limits the application of reciprocating pumps for many chemical and process services, although the type is not excluded from such duties where gland leakage can be accepted.

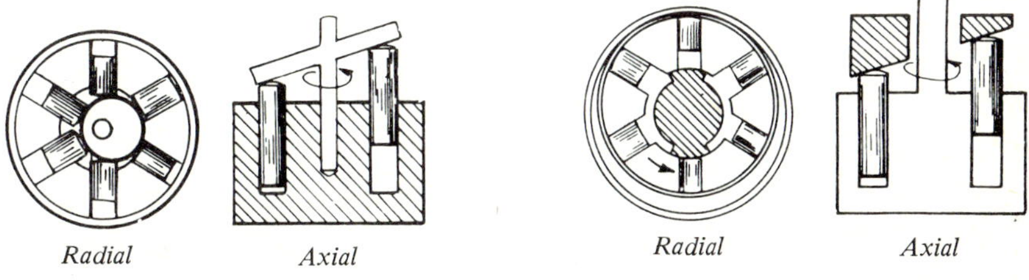

Rotating cam. *Fig 5.10* *Rotating block.*

Rotary Piston Pumps (see also Fig 5.10)

The so-called rotary-piston pumps are essentially multi-cylinder reciprocating pumps with a radial or axial arrangement of the cylinders – generally referred to as radial or axial piston pumps, respectively. In both cases the cylinder block may be fixed or stationary, with the pistons driven by a cam (radial pump) or a titled swashplate (axial pump); or the cylinder block may rotate as a whole round a fixed cam, or a fixed swashplate. The overall effect is the same in all cases, each piston being reciprocated in its respective cylinder once per revolution, with common feed to and delivery from the individual cylinders.

Such pumps have the same overall characteristics as multi-cylinder reciprocating pumps (which they are) and with five or seven (or more) pistons are capable of giving a substantially smooth delivery (although not pulse free). The layout makes for particularly compact construction using small diameter pistons which are usually of true plunger form (hence the common description rotary-plunger pump). Whilst displacement is somewhat limited as a consequence this can be partly offset by the use of higher running speeds, although more usually rotary-plunger pumps are specifically produced for high pressure services (*eg* hydraulic systems).

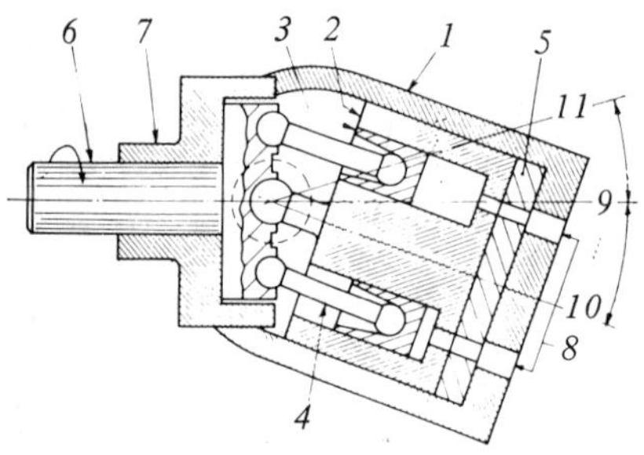

Variable delivery axial piston pump.

1. Outer casing mounted on trunnions. 2. Rotating barrel. 3. Piston. 4. Connecting rod. 5. Valve plate. 6. Main shaft. 7. Fixed housing. 8. Ports. 9. No delivery position. 10. Forward delivery position. 11. Reverse delivery position.

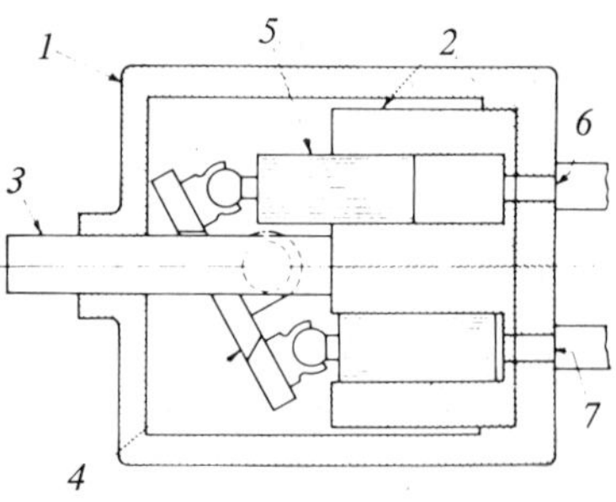

Axial plunger pump.

1. Pump casing. 2. Rotating barrel. 3. Driving shaft. 4. Swashplate. 5. Piston. 6. Inlet. 7. Outlet.

Axial Pumps

Axial type pumps also have a unique feature. Altering the angle of the swashplate alters the stroke of each piston or plunger and hence the displacement per revolution. Thus such pumps can be given variable delivery characteristics fully variable from maximum in one direction through zero to maximum in the opposite direction (using suitable porting) if required, simply by adjustment of swashplate angle, or by adjustment of cylinder block inclination if the swashplate is fixed. This relatively straight-forward method of achieving variable delivery can be of particular advantage in specific applications since it eliminates the need for any separate form of throttle or speed control and the pump can continue to run at a constant speed throughout.

Radial Pumps

Radial type pumps can also be given variable delivery characteristics by similar means to that employed on horizontal and vertical reciprocating pumps (*eg* variable throw eccentrics or other stroke-varying mechanisms). The range of flow control available, however, is not as complete as that possible with swashplate pumps.

Diaphragm Pumps

The diaphragm pump is a glandless type employing a flexible diaphragm which is oscillated in a shallow chamber to produce a similar pumping action to that of a reciprocating pump, but with a very much more limited stroke.

The particular advantage offered by a diaphragm pump is that only the diaphragm and the suction and discharge chamber are wetted by the fluid being handled. Also, although valves are still required the flow pattern is straightforward and simple ball valves can be used (although disc type valves or even flap valves may be used on some designs), yielding relatively open port areas. Thus diaphragm pumps are readily suitable for handling corrosive liquids, and also those containing a high proportion of solids, even abrasive solids. The basic design is inherently capable of passing the latter without damage or clogging, whilst corrosion resistance is simply a matter of material compatibility for the diaphragm itself and the pump chamber. The diaphragm material selected is normally a compatible elastomer (see **Table I**) whilst the pump chamber or liquid head can also be non-metallic, if necessary, with glass or ebonite ball valves.

TABLE I – TYPICAL DIAPHRAGM MATERIALS

Diaphragm	Service
'C' HYCAR	Mineral, animal and vegetable oils and fats. Petroleum hydrocarbons and solvents – petrol, paraffin, crude oil *etc*. Maximum temperature, depending on fluid handled 80°C
'B' BUTYL	Higher concentrations of acids not handled by other grades. Resistant to some solvents, but not full strength aromatic hydrocarbons. Maximum temperature under all conditions, 60°C
HYPALON (Chlorosulphonated Polyethylene)	Oxidising agents, mineral and vegetable oils. Maximum temperature up to 120°C intermittently, 100°C fairly indefinitely
PTFE (Faced) (Polytetrafluoroethylene)	The most versatile material, being attacked only by molten sodium and some fluorine compounds. Suitable for extremes of temperature according to the material being handled.

Other types of valves may be used, including flap valves, petal valves, *etc*, although these are normally confined to special applications or special designs. Ball valves are the more common for general purpose diaphragm pumps.

Double diaphragms may be used as a safety or back-up feature. In the former case the chamber between the two diaphragms is filled with an inert fluid so that in the event of failure of one diaphragm the product being handled is still contained within the pump (the pump can also continue to operate but without the safety feature). Mechanical diaphragm pumps may have two diaphragms operated synchronously in order to maintain pumping action in the event of failure of one diaphragm – *eg* a principle used on many i/c engine fuel pumps.

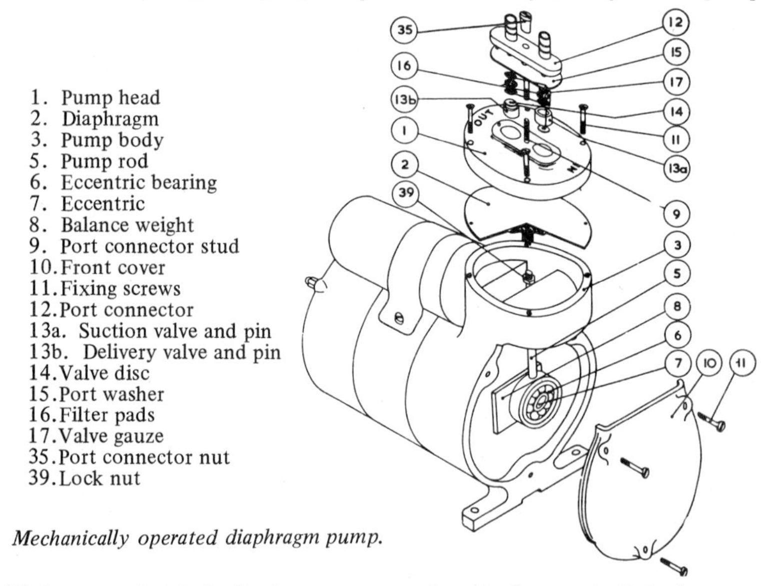

Mechanically operated diaphragm pump.

With a *mechanical diaphragm pump* the diaphragm is linked to a crank motion or similar mechanism to produce oscillation. The diaphragm is otherwise unsupported except around its periphery and thus one face is subject to full fluid pressure and the other to atmospheric pressure. This limits maximum pressure which can be generated with an elastomeric diaphragm to about 50 lb/in^2 (3.5 bar).

With a *piston-diaphragm* pump the chamber on the 'driving' side of the diaphragm is filled with fluid subject to pressure pulsations by a reciprocating piston or plunger, *ie* the diaphragm is oscillated by pulsating hydraulic pressure. Pressure on the diaphragm itself is the same on both faces, substantially reducing material stress and enabling much higher output pressures to be developed. Small diaphragm pumps of this type are widely used for metering and proportioning duties. Very much larger piston-diaphragm pumps are used for process work.

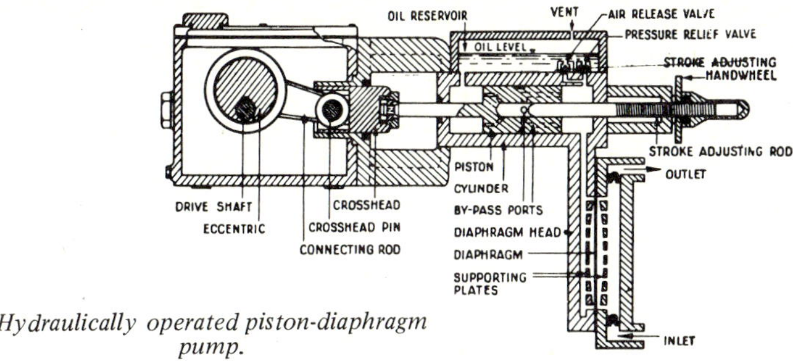

Hydraulically operated piston-diaphragm pump.

With an *air operated diaphragm* pump the diaphragm is reciprocated by compressed air. There are various ways of doing this. One is via compressed air supplied through a solenoid valve – see **Fig 5.11.** This valve alternately admits and releases compressed air at the pump head at a rate determined by the setting of a simple timer. Such pumps normally work under a positive suction head of at least 6 ft, this head being required to return the diaphragm ready for the next delivery stroke. If the air is vented to a vacuum system instead of the atmosphere, however, a positive suction head is not necessary and suction lifts can be accommodated.

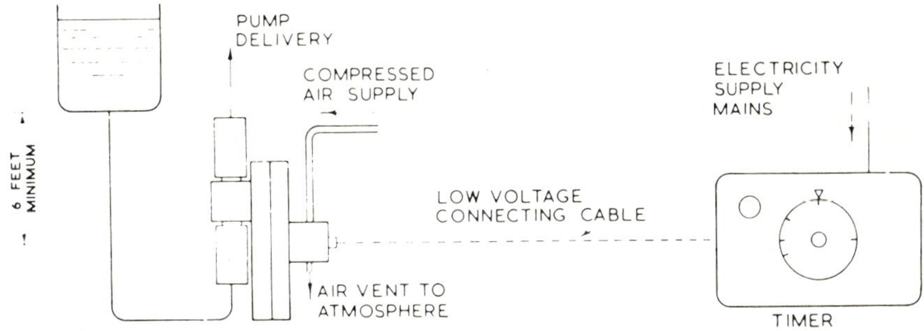

Fig 5.11: Air-operated diaphragm pump.

An alternative type of air driven diaphragm pump is the configuration where the diaphragm is hydraulically operated via a double-acting air motor. This is, in effect, a standard piston-diaphragm set-up, but using an air motor instead of a mechanically driven piston pump, although speed may still be controlled by a timer to achieve variable delivery (or more simply by a restricting valve in the air outlet). In this case the diaphragm pump will generate suction lift.

Another air-operated type is shown in **Fig 5.12** which consists of two identical diaphragm pumps connected by a pump rod, central stationary assembly and two product manifolds. Each of the pump ends has a pair of non-return valves

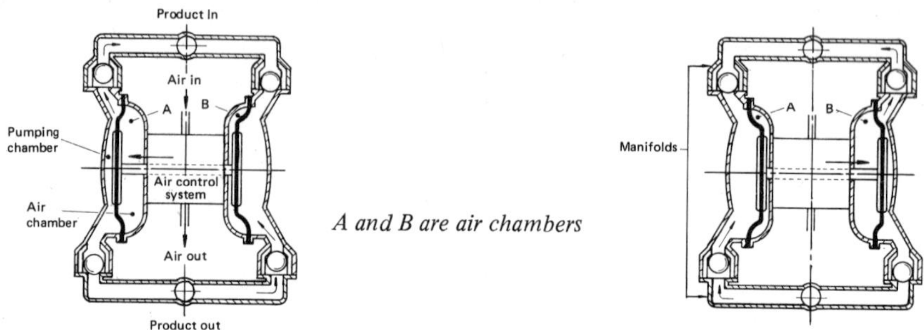

Fig 5.12: Selwood air-operated double diaphragm pump.

between the product chamber branches and the manifolds and an air chamber connecting to the central assembly. Switching of air to produce reciprocating motion is commonly achieved through a shuttle valve, the position of which is switched with reference to the pump rod position, either pneumatically or mechanically. It can be noted that with this type, although 'pressure balanced' the diaphragm is in fact subject to alternating pressure since air pressure will be higher than product pressure during the induction stroke.

There is also a further type of diaphragm pump where the diaphragm is driven by an electro-magnetic plunger either directly or indirectly (hydraulically), energised by pulses of electric current. Such pumps are produced in small sizes only as *dosing pumps.*

Characteristics

The displacement of a diaphragm pump is inherently limited since only a relatively short stroke can be employed, relative to the diaphragm diameter. Whilst a longer stroke may be geometrically possible this would simply over-stretch and thus over-stress the diaphragm material. The pressure which can be developed is also limited by the nature of the diaphragm material in the case of a mechanically driven diaphragm, although this is not necessarily the case with a piston-diaphragm pump since the diaphragm always operates with equal fluid pressure distributed uniformly over both faces of the diaphragm.

A diaphragm pump inherently produces a pulsating discharge although this can be cushioned by incorporating an air chamber on the delivery side, if necessary (*eg* where the pump is required to produce a substantially constant pressure output, such as for spraying). The H-Q curve will be substantially vertical with a definite cut off at the maximum rated pressure – **Fig 5.13.** Variable delivery can be provided by varying the stroke with constant speed drive. Maximum capacity is determined by the maximum permissible speed which is normally relatively low. Choice of a suitable size of pump, however, must take into account the head involved as if the pump is worked near its maximum permissible pressure

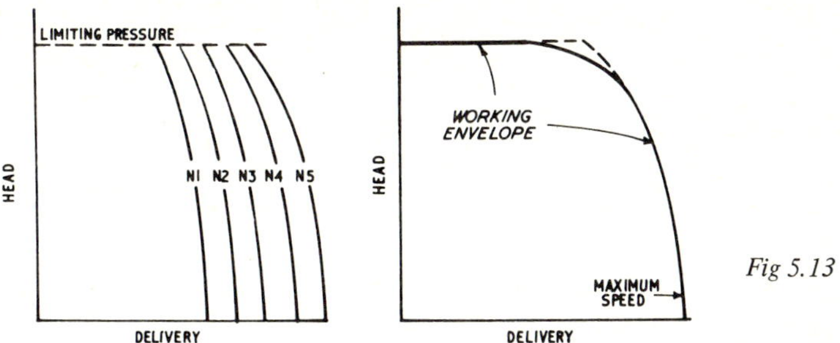

Fig 5.13

its effective capacity may be substantially reduced, depending on the detail design of the pump. Efficiency figures are never quoted for diaphragm pumps, but are usually low and of the order of only 10–25% in the case of mechanical diaphragm pumps.

Typically, a single-acting mechanical diaphragm pump delivering 5 gal/min (22 lit/min) of water would require about 1½ hp input for a working pressure of 40 lb/in^2 (3 bar) and about 2 hp input for a working pressure of 100 lb/in^2 (7 bar). Corresponding efficiency figures are approximately 10 and 20%, respectively. A double-acting diaphragm pump delivering 10 gal/min (44 lit/min) would require about 2 hp input for 40 lb/in^2 (3 bar) delivery, or 3 hp input for 100 lb/in^2 (7 bar) delivery, with efficiency figures of 15 and 25% approximately, respectively. The virtue of the diaphragm pump is in its general usefulness and basic simplicity rather than its overall efficiency.

Diaphragm pumps have good suction characteristics and thus are self-priming, indeed they usually operate most satisfactorily under conditions of suction lift, although priming may be needed for suction lifts exceeding about 15 ft (4.5 m). Diaphragm pumps can also operate as vacuum pumps as well as pressure pumps.

Diaphragm pump heads are relatively compact units and two or more can be connected in series or parallel with a common driver to provide higher pressures or capacities, respectively. With paralleled operation two diaphragms may be combined in a single head assembly to give a double-acting pump. Single- or double-acting pumps can also be connected for Duplex working. Larger pumps of this type may give capacities up to or exceeding 500 gal/min (2 250 lit/min) hydraulic operation being usual in such cases with an electric motor driver for the piston pump.

ROTARY PUMPS

Rotary pumps are basically machines with a rotating element or elements which squeeze or otherwise displace a liquid inducted into a casing to provide a substantially pulse-free delivery. They thus provide positive displacement characteristics, with discharge proportional to speed and theoretically independent of head – *ie* similar to reciprocating pumps, although the actual volumetric efficiency achieved may be lower. A marked difference, compared with reciprocating pumps, is that rotary pumps normally require no valves but only port openings. They are also considerably more restricted in the size range possibly, larger rotary pumps being the exception rather than the rule.

Originally, at least, rotary pumps were considered as viscous liquid pumps, with a particular suitability for handling oils – although the vane pump and the lobe rotor pump have also been used as gas compressors and exhausters since their conception. The considerable variety of types which have evolved, however, have extended the use of rotary pumps to a wide variety of general services and special duties where low to moderate deliveries are required, and where the fluid to be handled is substantially free from hard or abrasive contaminants.

Capacity of Slip

Theoretical capacity of a positive-displacement rotary pump is proportional to speed and independent of head. In practice, internal leakage will introduce losses which are more or less proportional to delivery pressure, so that the head-capacity curves for different driven speeds will not be truly vertical due to slip (see **Fig 5.14**). Slip will generally be quite low where the pump elements are of rigid construction but will usually tend to become more marked with increasing pressure with flexible rotor elements. In the latter case there will also be a limit to delivery pressure.

With rigid elements there is no theoretical upper limit to pressure developed other than the strength of the elements or casing involved, although in practice

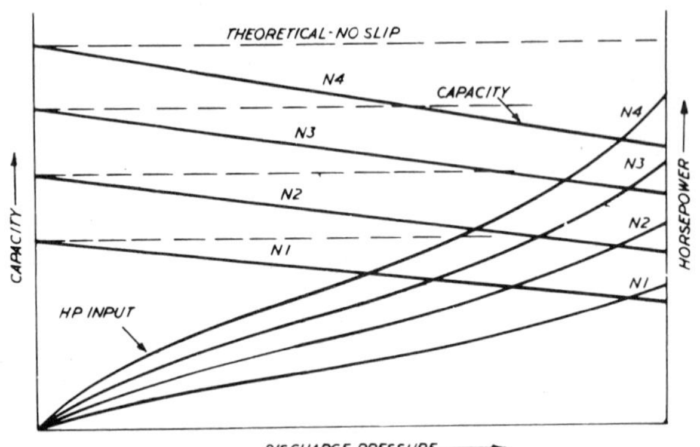

Fig 5.14: Typical performance of positive-displacement rotary pumps for capacity plotted against discharge pressure (or head) and corresponding horsepower input curves. Speeds N1, N2, N3 and N4 increasing.

high pressure working calls for considerable attention to detail design and accuracy of construction in order to avoid excessive slip and consequently marked loss of efficiency. In many types of pumps intended to operate against unlimited heads, a pressure relief valve may be built into the casing to guard against the possibility of abnormally high internal pressures being generated.

Efficiency

Efficiency can be derived only from the power input required. Thus, whilst increasing the running speed increases capacity more or less pro rata, the corresponding increase in horsepower required does not necessarily follow the same relationship. This will depend both on the design of the individual pump and on the viscosity of the fluid being handled. For a constant viscosity there is an optimum design speed at which pump efficiency will be a maximum, although this is usually completely non-critical with fluids of low to moderate viscosity. A rotary pump can therefore often be used at widely differing speeds as opposed to a nominal design speed without suffering any appreciable loss of efficiency. There is also the point that relatively small capacities are normally involved, so that the actual cost of any wasted power is usually negligible.

With increasing fluid viscosity the problem of efficiency can become more acute. Apart from the fact that this will inevitably increase the input power required to maintain a given capacity, efficiency will also tend to decrease with increasing viscosity due to modifications of flow conditions. This can even set a limit to capacity by establishing a critical speed at which flow breakdown or separation takes place because the fluid cannot flow into the casing rapidly enough to fill it. To some extent this may be countered by modification of the detail design or form of rotor but the more usual solution is to reduce the speed and thus the capacity available when handling a more viscous fluid. This, coupled with any desirable modification of rotor form, can maintain high pumping efficiency.

In theory, at least, this effect can be self-adjusting with a given type of rotary pump, as with a given horsepower available an increase in fluid viscosity will increase the horsepower required and automatically readjust the speed to a lower operating point. Unless the driver has a particularly flat BHP curve, and is also flexible enough to accommodate a speed change with increasing load, the real loss in speed and therefore capacity can be lower than anticipated **(Fig 5.15).** It is usually more realistic, therefore, to adjust either the drive speed gearing to provide an optimum pump speed for handling a viscous fluid and accept a reduced capacity whilst maintaining a similar overall efficiency figure; or to increase the power input to maintain a given required capacity. The former method is commonly adopted by pump manufacturers by giving a 'speed reduction' factor (or percentage) for handling more viscous fluids (usually only significant above a fluid viscosity of about 200 centistokes). Loss of capacity can be

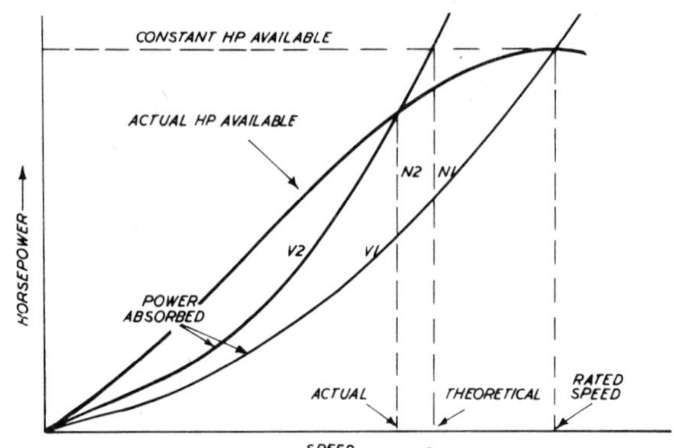

Fig 5.15: Horsepower absorbed curves for low viscosity fluid V1 on which pump performance is rated and with higher fluid viscosity V2.

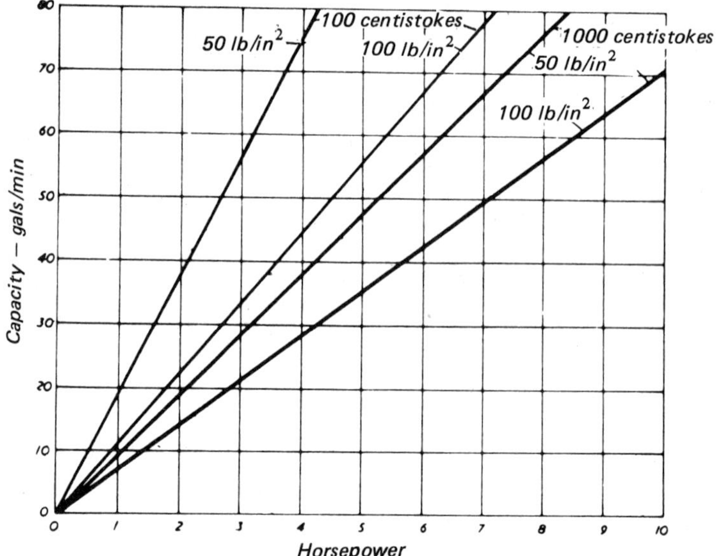

Fig 5.16: Horsepower performance of a typical screw pump plotted against delivery for two fluids at delivery heads of 50 and 100 lb/in² (3.5 and 7 bar). Increase in power input to maintain capacity can be determined by reading horizontally across and then vertically down.

estimated as identical to the speed reduction, with the pump requiring the same power input as the original rating. If capacity is to be maintained, this can be done only by increasing the power input. **Fig 5.16** shows comparative performance figures for a typical three-screw rotary pump.

Pumps for High Viscosities

With very high liquid viscosities it may be necessary to adopt special designs of rotary pumps (*eg* screw pumps) and low operating speeds in order to avoid channelling, *etc.* In such cases the only way to increase capacity is to increase the size of the pump so that although flow velocity is low flow rate can be quite high and produced at high efficiency. Standard rotary pumps, however, may be quite capable of handling fluids up to 10 000–20 000 centistokes viscosity with no more than 50% speed reduction and to even be workable at still further reduced speeds at viscosities up to 50 000 centistokes. Special designs are required to handle higher viscosities.

Priming Characteristics

The flexibility of operation of a rotary pump is further enhanced by the fact that it is usually self priming (provided some fluid is retained in the casing); and can, if necessary, handle entrained and dissolved gases or air. In the latter case the capacity of the pump is reduced in direct proportion to the amount of entrained gas, since this will tend to separate out within the pump and occupy a proportion of the pump internal volume. The actual capacity loss will normally be greater than the actual proportion of entrained or dissolved gas and increase with developing vacuum at the pump inlet. Under unfavourable conditions and with high vacuum at the inlet a capacity loss of 20–40% can be experienced. Such conditions can also lead to pump chatter and noisy operation, so that rotary pump applications involving the handling of fluids which may contain high proportions of entrained or dissolved gases or air may need special consideration. On the other hand, rotary pumps are widely – and successfully – used for handling petroleum fluids which inherently tend to contain a high proportion of dissolved air at normal atmospheric pressure and temperature. The main point to be appreciated in such applications is that a loss of capacity over rated performance may be anticipated as a consequence of the dissolved air present in the fluid.

It is a general characteristic of many types of positive rotary pumps, particularly the simpler designs, that they will pump in either direction with equal efficiency simply by changing the direction of rotation. Also, where the pumping action is derived from an eccentrically mounted rotor, variable delivery characteristics can often be achieved quite simply by geometric displacement of the casing or stator.

TABLE II – SUMMARY OF ROTARY PUMPS

Type	Capacity Characteristics	Head Characteristics	Suitability	Remarks	Practical Suction Lift
Simple Impeller	directly proportional to speed	low	low volume water pumping	relatively inefficient	5-10 ft (1.5-3 m)
Flexible Impeller	directly proportional to speed but decreases with increasing head up to 110 gal/min (490 lit/min)	limited delivery pressures; increasing delivery pressure reduces delivery at a given speed (40-50 lb/in^2) up to 125 bar	very versatile pump for handling clear or relatively clear fluids of low to moderate viscosity; excellent water pumps	small sizes only – must not be run dry; compact construction	up to 15 ft (5 m)
External Gear	directly proportional to speed	very little effect on capacity (up to 3000 lb/in^2) (210 bar)	clear fluids over a wide range of viscosities	speed and capacity generally reduced for handling higher fluid viscosities; must not be run dry	5-25 ft (1.5-8 m)
Internal Gear	directly proportional to speed (up to 500 gal/min)	very little effect on capacity (up to 200 ft)	clear fluids over a wide range of viscosities	speed reduced for handling higher fluid viscosities; must not be run dry	5 ft (1.5 m)
Lobe-Rotor	directly proportional to speed	little or no effect on capacity	clear fluids, especially higher viscosity fluids; with suitable rotor forms also suitable for handling gases		not self priming
Multi-Rotor	directly proportional to speed		special designs to suit particular applications, *etc,* also widely used as gas or air compressors	similar in principle to lobe rotor pumps, considerable variety of rotor forms; can run dry	not self priming
Single Screw	200-80000 gal/min (900 lit/min - 365 m^3/min)	25 ft (7.5m) per stage	water raising	large in size, slow running	none

cont...

TABLE II – SUMMARY OF ROTARY PUMPS (contd.)

Type	Capacity Characteristics	Head Characteristics	Suitability	Remarks	Practical Suction Lift
Rigid Screw (turn screw)	directly proportional to speed	low heads usual and low flow velocities; max pressure about 200 lb/in^2 (13.5 bar)	clear fluids only, but suitable for handling sensitive and very viscous fluids	wide range of sizes; also multiple screw arrangements	15-20 ft (5-6 m)
Eccentric Screw	directly proportional to speed	low heads usual and low flow velocities (150 ft (45 m) per stage).	readily adapted to handling difficult, sensitive or very viscous fluids; suitable for abrasive fluids		15-25 ft (5-8 m)
Sliding Vane	capacity increases with speed, but depends on internal leakage	increasing head tends to reduce capacity; usually poor sealing at low speeds	vacuum pumps and compressors; low to moderate pressure pumps for fluids	light, compact construction	6 ft (2 m) typical
Swinging Vane	directly proportional to speed	increasing head tends to decrease efficiency	handling non-viscous and viscous fluids with or without solid contaminants	larger sizes than sliding vane pumps	
Rolling Vane	directly proportional to speed		mainly used as a pump for spraying water		
Cam-Vane	directly proportional to speed (less slippage)		mainly used for vacuum pumps or gas compressors	unlike most rotary pumps may incorporate an exhaust valve	
Liquid Ring	directly proportional to speed		vacuum or gas pumping work, also compressors		up to 25 ft (7.5 m)
Roots	proportional to speed and backing pressure	limited compression ratio	blower or vacuum duties	air or gases only	

SIMPLE IMPELLER PUMPS

Simple impeller pumps are restricted to small sizes with limited capacities and may employ either a rigid or flexible impeller. A *rigid impeller* pump in this category is simply a small single-stage centrifugal pump which, because of its size, operates with a specific speed giving a low hydraulic efficiency. (Its efficiency will be directly related to its actual specific speed, as well as detail design). It is cheap and simple to produce, can operate on snore or running dry (since the impeller runs with clearance in the casing), and is a particularly attractive choice for low-volume, non-critical requirements involving clear fluids as the product being handled (usually water).

It is not self-priming and thus needs to be operated on flooded suction. A typical construction is a vertical assembly with the pump casing terminating in a stool with integrally mounted electric motor immediately above the pumping unit. Lift available from such pumps is usually strictly limited (*eg* possibly to less than 3 ft (1 m) in smaller sizes. The impeller can also be jammed by solids entering the pump chamber which can stall the motor causing windings to overheat. For this reason a screen or strainer is required on the suction side if this type of pump is used with dirty water, *etc.*

FLEXIBLE IMPELLER PUMPS

The *flexible impeller* pump is considerably more versatile although there are practical limits to the size in which it can be produced. It is also limited in the amount of pressure it can develop due to the deformable nature of the impeller. Sizes range from fractional horsepower units up to larger models with capacity range up to about 100 gal/min (450 lit/min).

Normally such pumps are designed for working at pressures of up to about 40–50 lb/in^2 (3 bar) maximum, although high pressure types are available for pressures up to 80–125 lb/in^2 (5.5–8.5 bar) maximum. Practical suction lifts of up to 15 ft (5 m) are possible.

The typical form of such a pump is a circular casing with inlet and outlet ports and piping housing an elastomeric impeller. The inner section of the casing is shaped to include an eccentric section across the two ports.

The working principle of the pump can be followed from **Fig 5.17**, *viz:*

(a) When the impeller leaves the eccentric section of the pump body, a vacuum is formed when the flexible impeller blades extend (the volume in the pump chambers increases) and, as a result of this, the liquid is sucked into the pump.

This gives the flexible vane pump instant self-priming characteristics with a high suction-lift potential. However it is not advisable to start up such a pump dry as initial dry rubbing could damage the elastomeric impeller tips. In normal use there is generally enough fluid retained in the casing to ensure wet-starting.

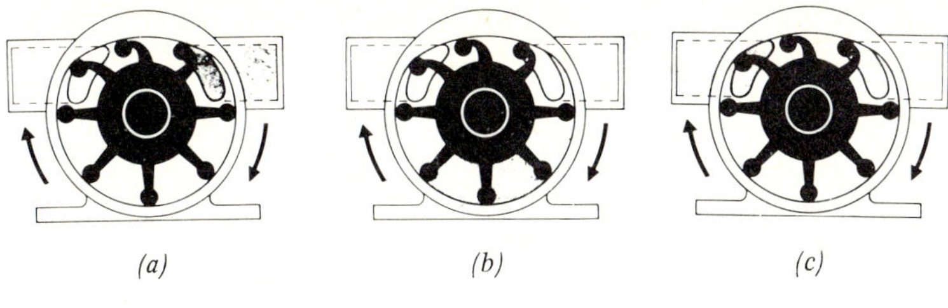

Fig 5.17

(b) As the impeller rotates, the fluid accompanies it from suction to the feed side – while the following blade sucks in more liquid. Since the impeller blades are not bent, this means that sensitive or easily damaged products can be pumped without any change in consistency and, in addition to this, comparatively large particles can pass through.

(c) When the flexible impeller blades come into contact with the eccentric part of the pump body, the blades are bent and the liquid is squeezed out of the pump in a smooth flow.

If the direction of rotation is reversed, pumping can be carried out in the other directions.

Capacity is directly proportional to speed, but also decreases substantially with increasing head – see **Fig 5.18.** For handling low viscosity fluids efficiency tends to increase with increasing speed and operating speeds up to 5 000 rev/min are common. Efficiency decreases rapidly with increasing liquid viscosity and the type is not basically suited for handling viscous fluids. It can, however, handle a very wide variety of non-viscous and low viscosity fluids, simply by choice of a suitable casing and shaft material, and compatible elastomer.

Pumps of this type are normally produced as integral units with or without a pedestal mount. In the absence of a mount the casing may incorporate lugs

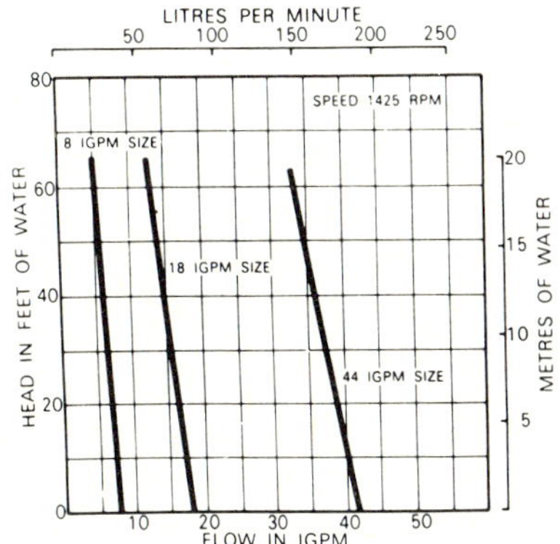

Fig 5.18: Typical H-Q curves for three sizes of flexible impeller pumps.

or an anchor plate for bolting directly to an electric motor or other fixing point (*eg* for the pump to be belt driven). One complete face of the casing is made detachable so that the impeller can be withdrawn and replaced as necessary.

Impellers are normally made of neoprene for general purpose applications. For handling oils, a nitrile rubber gives better compatibility; and for handling chemically active fluids a fluoroelastomer would be chosen – see **Table III**. The use of Viton will, however, normally reduce the pressure rating, and also the self-priming ability.

TABLE III – ELASTOMERS FOR FLEXIBLE IMPELLERS

Impeller Material	Application/s	Working Temperature Range °F	°C
Neoprene or Butyl	General purpose pumps	45–180	7–80
Nitrile	Maximum resistance to oils	45–180	7–80
Polyurethane	Maximum resistance to wear and abrasion	40–200	5–90
Fluoroelastomer (*eg* Viton)	Chemical solutions, solvents, *etc*	45–200	7–90

The fluid temperature which can be handled by a flexible vane pump is, of course, limited by the maximum service temperature of the elastomer employed for the impeller.

Since the impeller is flexible it has the ability to over-ride solids in suspension, so flexible vane pumps can be used for handling contaminated fluids. In this case a polyurethane impeller would normally be employed for improved resistance to abrasion and wear.

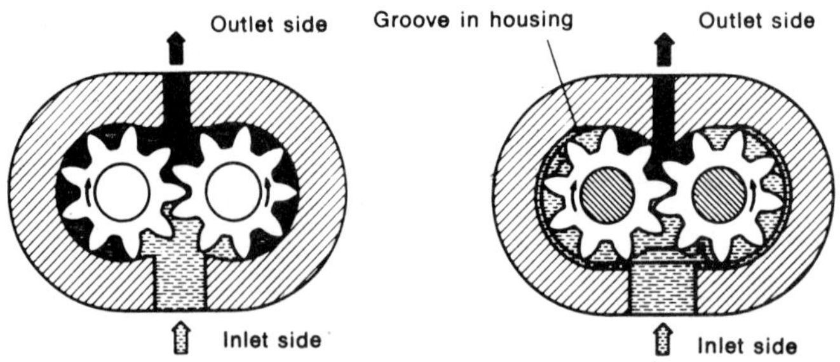

Fig 5.19: Simple external gear pump (left) and pressure-compensated type (right).

GEAR PUMPS (Fig 5.19)

Gear pumps are also suitable for high pressure working (*eg* up to 3 000 lb/in^2 or 210 bar), especially with oil fluids. This, however, places a premium on detail fluids). They are normally of all-metal construction and thus have limited suitability for handling corrosive fluids.

Gear pumps are also suitable for high pressure working (*eg* up to 3 000 lb/in^2), especially with oil fluids. This, however, places a premium on detail design, precision workmanship and rigidity and calls for particular attention to reduce internal leakage to a minimum, as well as attention to bearing design to resist torque and pressure loads. Efficiencies in excess of 90% are possible, although the figure achieved is generally lower with general purpose gear pumps.

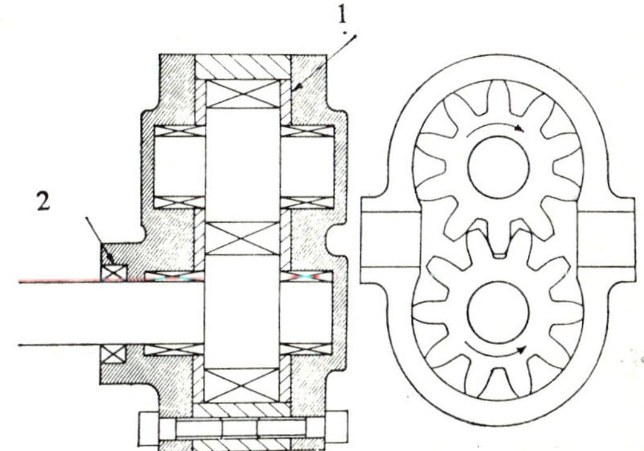

Fig 5.20: External gear pump.
1. Side plates (wear plates)
2. Shaft seal

External Gear Pumps (Fig 5.20)

External gear pumps are the simplest in design, consisting of two intermeshing gears of the same diameter and form mounted on separate spindles, one gear shaft being driven whilst the other idles. This type is readily produced in a wide variety of sizes. Sizes may range from miniature units up to large models with capacities up to 100 gal/min (450 lit/min) or more. It is a general characteristic of larger sizes that the decrease in capacity with increasing delivery pressure tends to be greater than with small sizes, although this is not invariably the case.

Pumping action is produced by nature of the fact that during rotation, as each pair of teeth intermesh on the inlet side, the volume on that side is reduced by the volume of two tooth spaces, providing a suction effect. Oil flowing into the suction space is then trapped on each side by a tooth crest approaching the bore of the housing and carried round to the delivery side by the 'pockets' between adjacent pairs of teeth. On the delivery side, the oil is displaced from the delivery port under pressure.

For maximum volumetric efficiency there should be no leakage between the teeth, and no leakage across the end faces of the gears. Even if such internal leakages are reduced to a practical minimum, there may be a further penalty to pay in the high bearing loads and unbalancing loads introduced by fluid trapped in the teeth pockets, which can result in increased friction and lowered mechanical efficiency. Thus a compromise may have to be reached in order to achieve maximum overall efficiency, although usually attention to increasing volumetric efficiency results in higher gains than any corresponding reduction in mechanical efficiency.

Capacity of a gear pump is normally expressed in terms of displacement per revolution. Accurate determination of displacement for a pair of gears is only possible when the method and extent of tooth venting at the meshing point is known but normally approximates closely to

$$Qd = 0.7854\,(D^2 - C^2) - 1.3\left(\frac{B}{N}\right)^2$$

where

D = outside diameter of gears

C = centre distance of gears

B = base circle diameter of gears

N = number of teeth on each gear

Qd is then the displacement per revolution per unit face width of the gears, in consistent units.

Internal Gear Pumps

The internal gear pump employs a smaller power driven gear mounted eccentrically with respect to a larger external idler gear contained in a circular casing. Rotation produces a sequence of expanding and contracting pockets into which liquid is drawn from the inlet side, transferred to the other side and forced out. The action is more gentle than that produced by an external gear pump, the generation of localised high fluid pressures is avoided and shearing forces generated on the fluid are reduced.

Individual designs may vary in detail. Thus with simple gear forms the necessary blocking action to prevent backflow from the outlet to the inlet side may be provided by the gears themselves, or by a crescent shaped partition fitted to the casing between the gears – **Fig 5.21.** With some designs the same pump may be capable of handling a wide range of liquid viscosities up to molasses, tar and bitumen with a reduction in speed (and thus capacity) to maintain efficient working with higher viscosities. With others the number of teeth may be reduced and shape of the teeth considerably modified for higher viscosity liquids as a means of maintaining capacity and efficiency.

The internal gear pump is more complicated, and thus more costly to produce, than the external gear pump. It is, however, better suited to handling

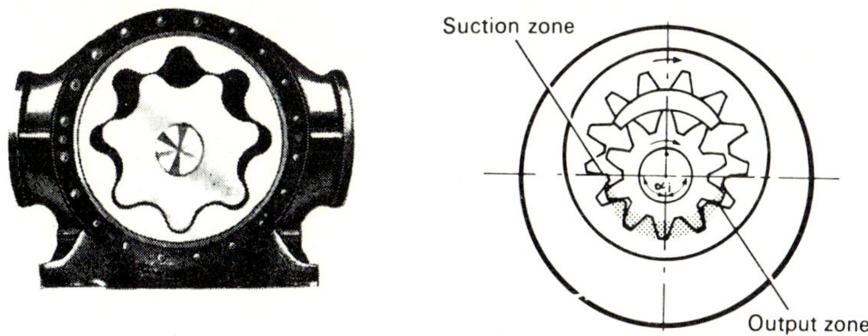

Fig 5.21 Internal gear pump.

liquids with higher viscosities and shear-sensitive fluids, and is thus basically employed for such services rather than as a general purpose pump. It may, however, be equally suitable for lower viscosity fluids, including volatile liquids, with suitable shaft seals. Like the external gear pump, since the rotating elements work with rubbing contact the type is not suitable for handling contaminated fluids, and it is desirable that the fluid should also be a lubricant, or the bearings of a type which are adequately lubricated by the fluid. If necessary, the bearings for the two gears can be mounted externally, with separate lubrication.

An advantage of the internal gear pump compared with an external gear pump is that it can produce larger displacements for the same overall size.

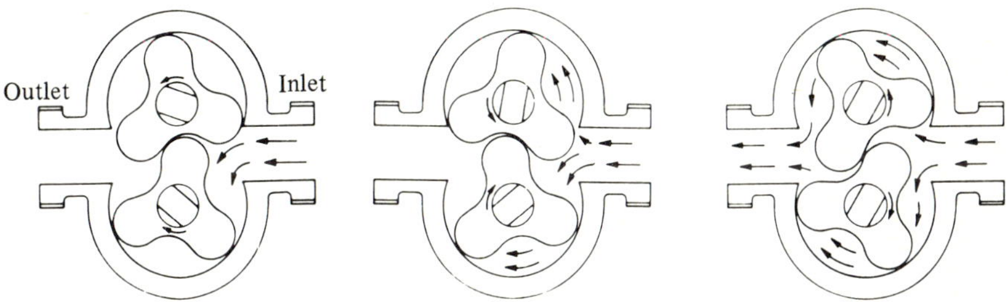

Fig 5.22: Operating principle of typical lobe rotor pump.

LOBE-ROTOR PUMPS (see also **Fig 5.22**)

Lobe-rotor pumps are used in all branches of industry. In the general food and beverage producing industries, as well as in pharamaceuticals production, their hygienic qualities are highly valued. Industrially, their high efficiency, robustness, reliability and potential for handling corrosive and often aggressive chemicals are particularly attractive.

Lobe-rotor pumps operate on a similar principle to internal gear pumps except that the opposite rotating intermeshing elements are of lobe rather than

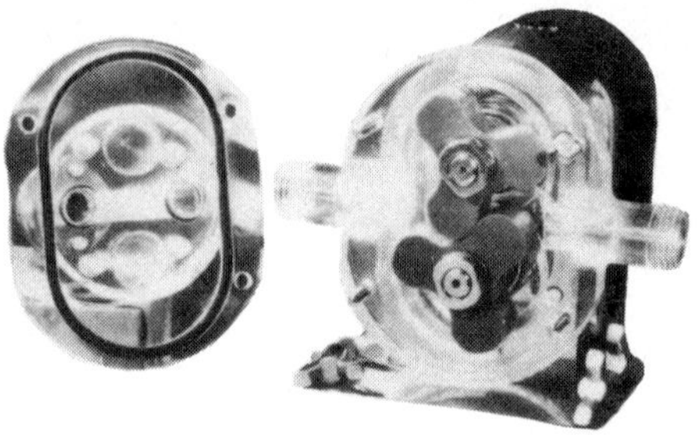

Lobe rotor pump.

gear form and maintain a small but positive clearance at all times. Both rotors are power driven and synchronised by interconnection gears or a timing chain mounted externally. There is thus no rubbing wear on the rotors themselves and if necessary the clearances can be adjusted to suit the fluid being handled, with particular respect to slip and operating efficiency; or even to pass solids. The considerable variety of rotor forms which can be employed also provides control over efficiency, suitability for handling a particular product, and capacity. Lobe rotor pumps are also suitable for handling gases as well as liquids, and are thus also used as compressors and exhausters.

Plain rotor forms are generally suitable for handling a wide variety of fluid viscosities, two-lobe, three or four-lobe rotors generally being preferred for straightforward duties. Displacement for any pair of rotors can be calculated from

$$\text{Displacement} = K(D^2 - C^2) \times W \times \text{rev/min}$$

where

D = rotor outside diameter

C = rotor centre distance

W = rotor width

K = is a coefficient determined by the rotor form and the clearances involved.

Empirically determined values of K are used by the pump designer to anticipate the performance of larger or smaller pumps using the same rotor form, but K values for different rotors does not necessarily afford a comparison of respective efficiencies.

Capacity Parameters

Whilst capacity is directly proportional to speed, it is rather more dependent on head than in the case of most other positive rotary pumps. For high head working 'recovery' of capacity by increasing the face width of the rotors considerably increases the load on the pump components and so normal practice is to decrease face width for high head lobe rotor pumps and go to a larger rotor diameter, if only to reduce the overhung load on the bearings.

In practice, capacity is usually adjusted by varying the length (face width) of the rotors – *ie* to produce a range of lobe rotor pumps of different capacities the same rotor diameter (and form favoured by the individual manufacturer) is retained, but with different rotor (and casing) lengths. Within this range clearances may then be adjusted to suit different fluid viscosities. Thus to achieve low slip with low viscosity, fluids clearances are reduced to a practical minimum (consistent with differential thermal expansion or physical distortion likely under working conditions). With increasing liquid viscosity slip will decrease, so clearances can be increased to reduce fluid shear and power input. Increasing clearances also enables speed to be maintained without developing excessive noise or vibration. Speed reduction may, however, be inevitable with increasing liquid viscosity in order to avoid flow breakdown or cavitation.

Pump Selection

Selection of a suitable lobe rotor pump, is not necessarily as straightforward as with other rotary types, since the optimum working range may be more restricted. Thus, basically, for optimum working a particular lobe rotor pump has a specific envelope defining its capacity and head range, with capacity determined by size and speed, and head by the rotor diameter:width ratio. Logically, selection of a suitable model should be based on the required working point coming within that specific envelope, as with centrifugal pumps. Requirements may be further modified by the liquid viscosity (*ie* requiring different clearances to standard). The nature of the fluid to be pumped may dictate the most favourable rotor form.

Products Handled

As with gear pumps there is generally some advantage in reducing the number of rotor lobes with increasing fluid viscosity. Thus alternative rotor forms may be provided for a particular design – say three or four lobe rotors for lower viscosity fluids and two lobe rotors for high viscosity liquids. This factory performance handling non-viscous fluids and gases. On the other hand, for handling fluids containing soft solids it is generally desirable to reduce the number of lobes on the rotor and employ simple forms to minimise the chances of the solids jamming in the pump. The shape of the rotor may be further modified in such cases to promote the most favourable pumping action, although this may present conflicting requirements. Thus decreasing the number of lobes will normally increase the cyclic pressure variations or pulsations, whereas a delicate product may require substantially pulse free transport through the pump.

Lobe rotor pumps are also capable of handling fluids containing entrained or dissolved gases, although there will be some loss of efficiency in such cases. Conventional rotors are not suitable for handling hard or abrasive solids in suspension, although coated or elastomeric rotors may be employed for such duties.

Priming Characteristics

Lobe rotor pumps are not self-priming and thus require operating with a positive suction head. Good suction lift is attainable once primed and so a convenient method of installation is often to locate the pump at the bottom of a U-bend in the suction line, or fit a non-return valve in the suction line when the pump is not operated on flooded suction.

SCREW PUMPS

Screw pumps fall into three distinct categories:

(i) Single-screw pumps (Archimedean screws).

(ii) Intermeshing screw pumps with two or more rigid screw-form rotors. These are also known as rigid screw pumps.

(iii) Single-screw pumps where a rigid (screw form) rotor rotates in a flexible stator. This type is also known variously as an *eccentric screw pump, eccentric worm pump* or *progressing cavity pump.*

Archimedean Screw

The *Archimedean screw* pump is normally produced in large sizes (*eg* from about 12 in (300 mm) diameter upwards) and in lengths up to about 50 ft (15 m). It is widely used as a water-raising pump with the screw arranged at an angle of about 30° – see **Fig 5.23.**

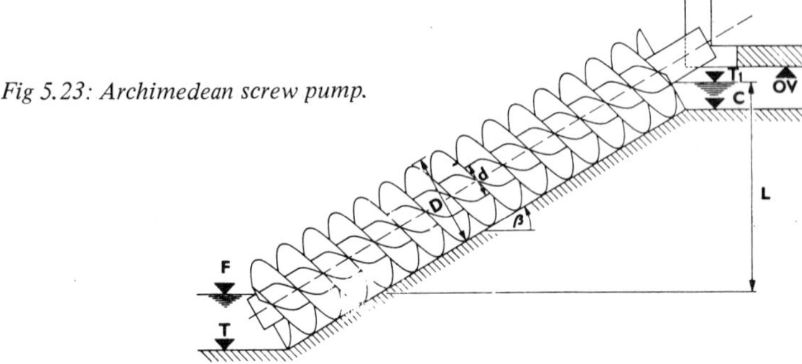

Fig 5.23: Archimedean screw pump.

D = *Screw diameter.*
d = *Centre tube diameter.*
β = *Angle of inclination.*
F = *Filling point, at which the screw works at its maximum capacity.*
T = *Touch point, at which the capacity is reduced to zero. For water levels between F and T the capacity of the screw pump adapts itself to the supply of water.*
C = *Chute point, at which the screw trough passes into the receiving channel.*
T_1 = *Maximum pumping point, to which the water is lifted for maximum efficiency.*
L = *Lift of screw.*

The same type of pump can also be used for handling liquids containing solids in suspension with either vertical lift or horizontal transport. The chief disadvantage of the Archimedean screw pump is the considerable bulk necessary to achieve high capacities since rotational speeds are usually of the order of 30–60 rev/min only.

The same principle has recently been introduced in very small pumps with the screw formed from circular or flat section wire rotating in a plastic tube.

Intermeshing Screw Pumps

In an intermeshing screw pump the screws may be in meshing contact, one power driven screw driving the other; or run with positive clearance, both screws being power driven. The most usual type is a two-screw pump with both screws being driven and operating with positive clearance. This may have a single pair of screws, in which case the axial thrust developed has to be taken by screw bearings, or two pairs of back-to-back screws where end thrust are balanced – **Fig 5.24.**

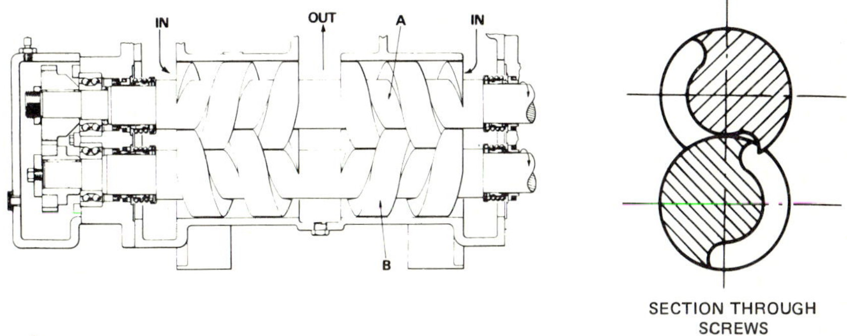

Fig 5.24: Section of typical external two-screw pump.

Delivery from a rigid screw pump is free from pulsations and there is no unswept volume (*ie* the whole of the pumping chamber is fully displaced). Maximum pressure is limited to about 200 lb/in^2 (13.5 bar) and maximum fluid viscosities which can be handled at normal speeds is of the order of 1 500 centistokes. Screw pumps are therefore suitable for handling a wide range of liquids from water to thick oils, *etc.* They can also be used to pump gases and air.

Rigid-screw pumps are particularly suitable for construction in a wide range of sizes, and can be run at high speeds, if necessary. Thus capacities available range from 1 gal/min or less up to 2 000 gal/min (9 000 lit/min) or more, from standard production models. Larger screw pumps are particularly used for the bulk handling of oils and similar fluids, but the basic type is suitable for handling most clean fluids with low flow velocities and at low heads.

Sectional view of a Mirrlees screw pump.

All screw pumps can also be operated at low speeds whilst maintaining good efficiency which, together with the low flow velocities induced, makes them suitable for handling viscous fluids and shear-sensitive fluids, with suitable modification of screw form, if necessary. Designs which work with positive clearance can accommodate contaminated fluids, whilst protection of the working parts can be given by hard facing or coating or lining and bearings and timing gear can be external and independently lubricated.

Screw pumps are generally self-priming when wet, although the amount of suction developed will depend on the clearances in the pump and also the viscosity of the fluid.

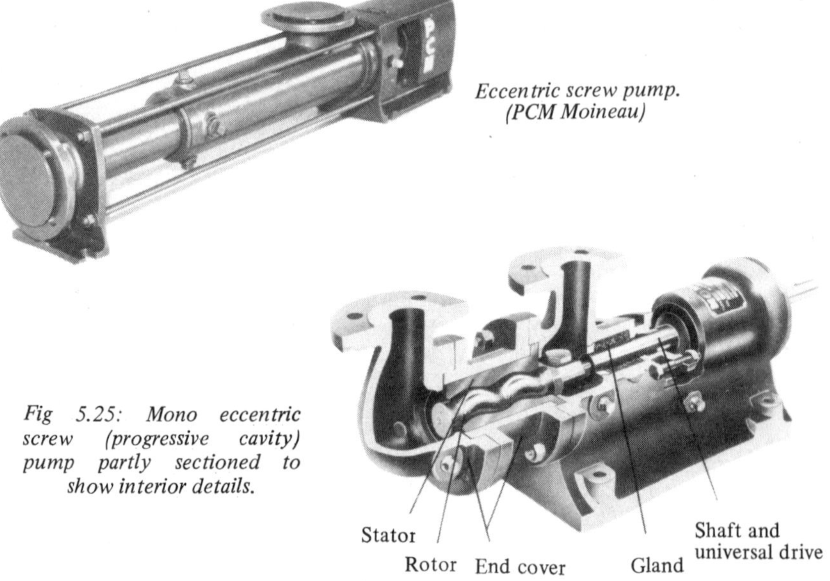

Eccentric screw pump. (PCM Moineau)

Fig 5.25: Mono eccentric screw (progressive cavity) pump partly sectioned to show interior details.

Eccentric Screw Pumps

The eccentric screw pump is a particularly versatile type capable of handling a variety of liquids and products with a gentle pumping action with high efficiency. It comprises, basically, a rigid screw-form rotor rolling in a resilient internal helical stator of hard or soft rubber with a moderately eccentric motion – **Fig 5.25**. The form originated with the Mono pump in 1936, actual design principles now employed varying in detail with the manufacturers concerned – *eg* see **Fig 5.26**.

The eccentric screw pump can be specifically tailored to handle viscous liquids, slurries, pastes, solids in suspension and even delicate products, largely because of the low flow velocities realised through the pump. Heads of up to 150 ft (45 m) are practical for each stage (*ie* each full pitch rotor length). Multi-stage pumps of this type are used for deepwell services.

Eccentric screw pumps are now available in capacity size up to 900 gal/min (250 m^3/bar).

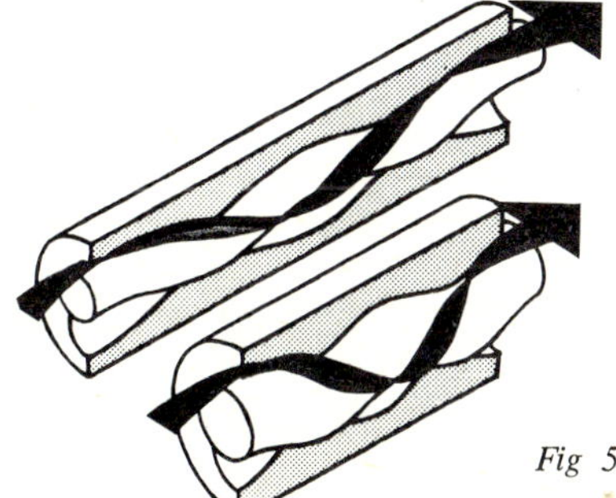

Fig 5.26: Variations in eccentric screw forms.

VANE PUMPS

Vane pumps may be considered under three separate categories (see also **Table IV**).

(i) Sliding-vane pumps – primarily high speed units for small to moderate capacities handling non-viscous and low viscosity fluids. Larger machines of this type are also produced for working as air (or gas) compressors.

(ii) Heavy-duty vane pumps – large, slow running machines capable of handling the most viscous fluids and semi-solids.

(iii) Swinging-vane pumps – medium or larger capacity units of rugged construction capable of handling contaminated fluids.

The simple sliding-vane pump comprises a slotted rotor mounted asymmetrically in a substantially circular casing. Rigid blades are fitted in the rotor slots and are free to slide radially – **Fig 5.27**. They are normally thrown outwards under the centrifugal force developed by rotation, although this radial

TABLE IV – APPLICATIONS OF VANE PUMPS

Application or Service	Type of Vane Pump	Remarks
Air or gas	Sliding vane	Laminated plastic vanes oil-sealed for vacuum services
Clean fluids and oils, low viscosity	Sliding vane	Metal or plastic vanes
Low viscosity fluids, large volumes, some contaminants	Swinging vane	Cast iron or bronze
Dirty water	Swinging vane	Cast iron
Liquids with solid contaminants	Swinging vane, heavy-duty vane	Cast iron or bronze
Viscous liquids	Sliding vane, heavy-duty vane	With spring loaded, cam or push-rod operated vanes
Cleaning liquid transfer	Sliding vane	Plastic vanes
Water spray (irrigation)	Rolling vane	
Heavily contaminated fluids, abrasive solids, semi-solids	Heavy-duty vane	Single full-diameter vane, hardened materials

Fixed capacity vane pump.

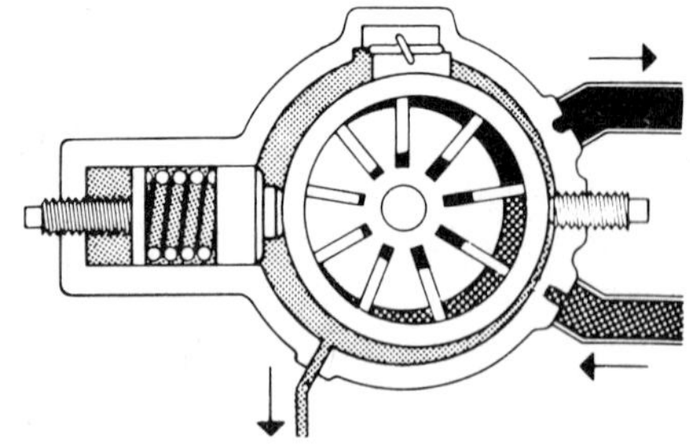

Variable capacity vane pump.

Fig 5.27

movement may be assisted by spring loading or pressurising the base of each vane. Sealing is provided by the vanes rubbing along the inner surface of the casing, the space between adjacent blades forming pockets which alternately expand and contract in volume. The inlet port is located at a point approximating to maximum pocket volume, the resulting depression sucking in fluid which is then carried round in the pocket until the volume is contracted again and the pocket contents squeezed out through the delivery port.

The optimum number of blades depends on the fluid viscosity, whilst the actual shape of the casing must be designed to provide uniform acceleration and deceleration of the blades as well as avoiding trapping to achieve maximum efficiency. In general, a large number of blades is used in a vane pump designed for handling gases (*eg* compressors and exhausters) and a smaller number (typically six or eight) for low viscosity liquids. For handling higher viscosity liquids it is advantageous to decrease the number of blades still further; and it may also be necessary to reduce the operating speed.

The basic vane pump principle lends itself to two-cell and multi-cell configuration, which has the immediate advantage of increasing the capacity for a given size, reducing the blade acceleration forces, reducing the sealing arcs and relieving the bearing loads. It is generally necessary, however, to increase the number of blades and so such pumps are little developed, except for handling non-viscous fluids **(Fig 5.28).**

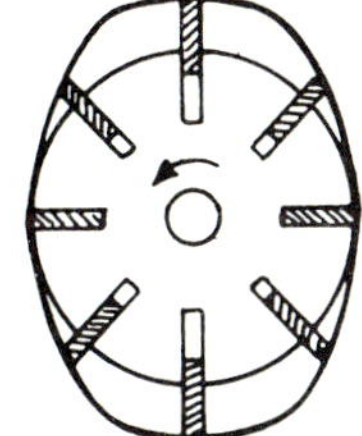

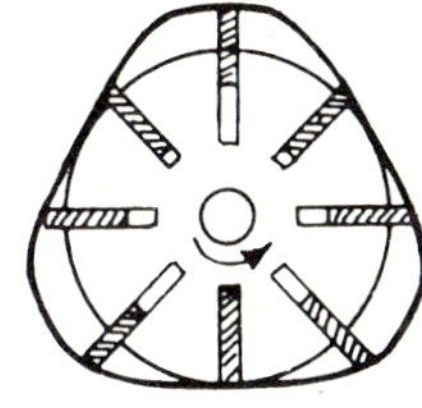

Fig 5.28: Two- and three-cell vane pumps. (Diagrammatic)

Rotor assembly of sliding vane air compressor.

Main application of sliding vane pumps ia as rotary compressors or exhausters. They are, however, suitable for low to moderate pressure pumps for handling non-viscous fluids, and oil fluids; also low pressure hydraulics. They are not suitable for use with contaminated fluids. It is advantageous if the fluid is a lubricant, although this is not necessary with a suitable choice of vane material (*eg* a laminated plastic). A particular attraction is that a vane pump is light and compact, and also relatively simple in construction. Without mechanical assistance to extend the vanes the pump will cease working and will not self-prime at a certain minimum speed (typically about 500 rev/min).

Heavy-Duty Vane Pump

The heavy-duty vane pump works on exactly the same principle except that both the casing and rotor construction is extremely robust, with solid metal vanes reduced in number to three, two or even one.

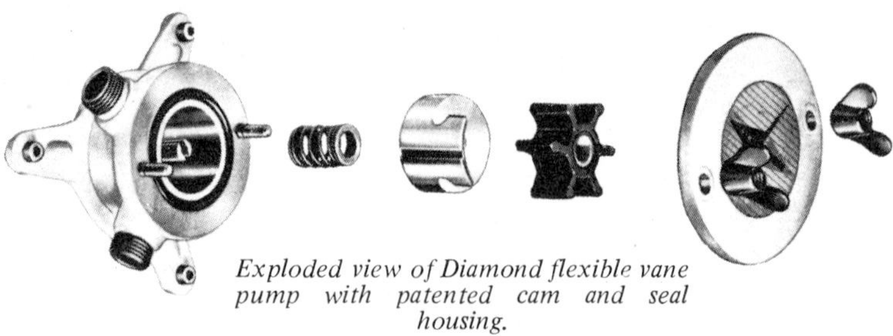

Exploded view of Diamond flexible vane pump with patented cam and seal housing.

Fig 5.29: Heavy-duty single-vane pump.

Previously such pumps employed spring-loaded vanes to ensure proper extension at slow rotational speeds, although this had obvious limitations. A considerable improvement is realised where a single vane is employed of full diameter with a perfectly circular casing, the rotor being mounted eccentrically and slotted to accommodate the vane which then merely slides from side to side, relative to the rotor – **Fig 5.29.**

Rotational speed can be very slow, depending on the viscosity of the product being handled, and may be as low as 1 rev/min handling semi-solids. Such pumps are capable of moving the most viscous liquids and liquid-solid mixtures (which may be 'pre-digested', if necessary to pass through the inlet port). Abrasive solids may be handled by such pumps using hardened materials, whilst the relatively gentle pumping action and low speed of transport makes the type also suitable for handling shear-sensitive mixtures.

Swinging Vane Pump

The swinging vane pump is also of rugged construction, the vanes being of cast iron or bronze, pivotally mounted on the rotor – **Fig 5.30.** Individual vanes can thus override solid contaminants present in the fluid. Production of the type is somewhat limited, but it is generally suitable for handling all types of low viscosity liquids compatible with the pump materials, and in particular dirty water and liquids with solid contaminants.

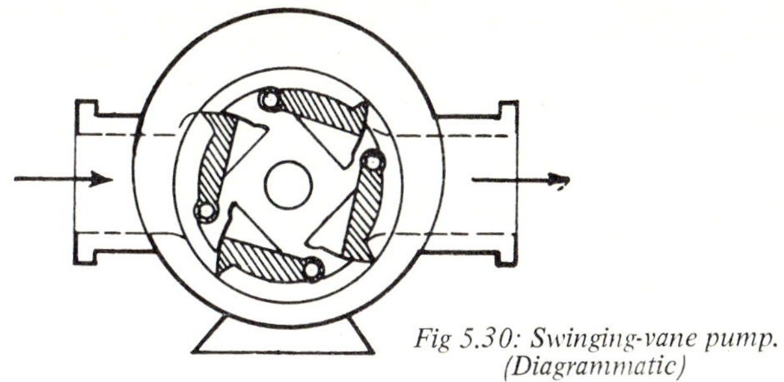

Fig 5.30: Swinging-vane pump. (Diagrammatic)

Fig 5.31: 'Genevac' cam-vane pump.

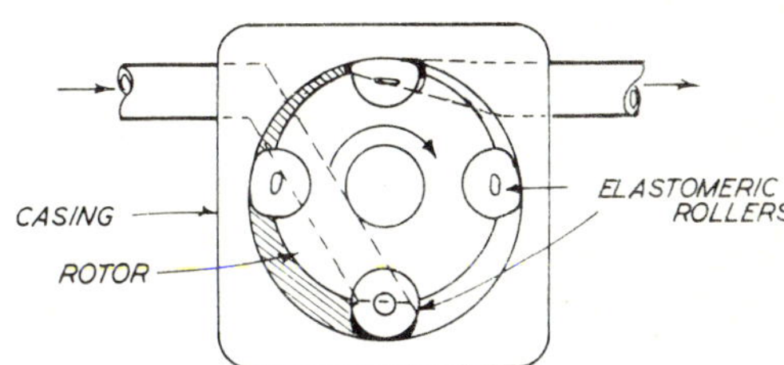

Fig 5.32: Rolling vane pump.

Vane Pump Variations

A number of variations on the vane pump principle have been developed, aimed principally at reducing or eliminating scraping surfaces and thus frictional wear. The cam-vane pump is one example where movement of a sliding vane is initiated by a cam. A patented development of this principle is shown in **Fig 5.31**, which is essentially designed as a vacuum pump.

Another variation is the rolling-vane pump, shown in **Fig 5.32**, where the rotor is grooved rather than slotted and the vanes are replaced by elastomeric sealing cylinders free to roll as well as slide.

LIQUID RING PUMPS

The application of liquid ring pumps is confined almost exclusively to the pumping of gases and vapours, either as vacuum pumps or low pressure compressors (blowers).

The basic form of a liquid ring pump consists of a multi-bladed rigid impeller eccentrically mounted within a circular casing which is partially filled with a liquid. Rotation of the impeller forces the liquid outwards against the periphery of the casing, creating a ring of liquid – hence the name. The pumping action provided is similar to that of a vane pump.

The impeller creates a series of pockets of cells between the individual vanes, but due to the eccentric position of the impeller the individual cell volumes undergo a change from a minimum to a maximum on each revolution – **Fig 5.33.** Ports cut in appropriate positions in the back of the casing will then provide suction entry on one side (in the region of expanding cell volume); and discharge under pressure on the other side (in the region of contracting cell volume). Thus the machine performs as a pump for the transfer of any fluid not miscible with the filling liquid. In practice, the filling fluid used is normally water (and this type of pump is commonly called a water ring pump); and the fluid pumped is a gas (normally air).

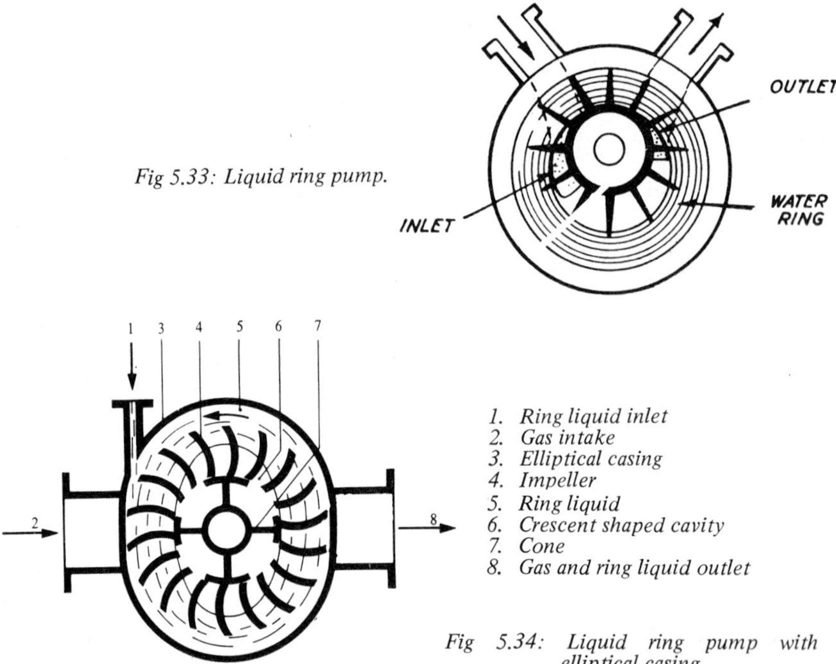

Fig 5.33: Liquid ring pump.

Fig 5.34: Liquid ring pump with elliptical casing.

If the casing is made elliptic instead of circular a two-cell or double-acting pump is produced, with the further advantage of radial forces on the impeller and balanced. This configuration is usually preferred for compressor duties – see **Fig 5.34.** A third type of liquid ring pump employs a screw form of impeller

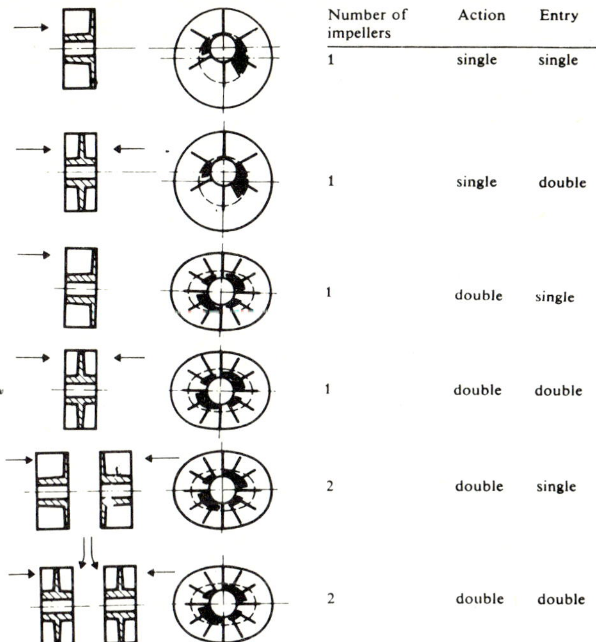

Number of impellers	Action	Entry
1	single	single
1	single	double
1	double	single
1	double	double
2	double	single
2	double	double

Fig 5.35: Types of liquid ring vacuum pumps.

in a circular casing, but has more limited application. Multiple-cell configurations are also possible, but increase the difficulty in maintaining the fluid ring. (See also **Fig 5.35**).

All liquid ring pumps have a minimum speed below which the liquid ring itself is not properly formed and so the pump will not work. For vacuum pumping duties the optimum circumferential speed range for a liquid ring pump is of the order of 50–60 ft/sec (17–20 m/sec). For compressor duties optimum circumferential speed range is of the order of 80–95 ft/sec (25–30 m/sec). Single stage pumps are capable of producing pressures up to about 115 lb/in^2 (8 bar); two-stage pumps up to 150 lb/in^2 (10 bar); and multi-stage pumps up to about 290 lb/in^2 (20 bar). Special separators are normally necessary to separate the delivered (compressed) gas from the ring liquid on the delivery side.

The compression ratio attained depends on the pressure in the rotating liquid ring and the shape of this ring, and also the size and shape of the ports. The liquid ring shape is not truly annular, due to the eccentric position of the impeller and consequently the different depth of immersion of the blades in the ring. The casing shape is sometimes modified from that of a true circle in order to provide a more consistent thickness of liquid ring.

Delivery

The theoretical discharge achievable with a liquid ring pump is given by the following formula:

$$Q\text{ (theoretical)} = \frac{t_m - s}{2t_m} \cdot w \cdot [(r_a - a)^2 - r_i^2]$$

t_m = mean pitch of cell filled with air
s = blade thickness
w = angular velocity
r_a = outer blade radius
a = immersion depth
r_i = inner blade radius

The actual discharge will be less than this by the amount of internal losses which can only be determined empirically. These losses increase as the pressure ratio between intake and discharge sides increase.

Overall efficiency is generally low (of the order of 25–35%), but can be improved slightly by the use of blades with small forward curvature.

Liquid ring pumps have a number of advantages for vacuum work in that they require little or no supervision or maintenance and can also accommodate slight impurities such as dust or small quantities of other fluids entrained with the gas being pumped.

Used as a compressor the water ring pump has the advantage of providing an oil-free delivery, although the compressed air may be heavily saturated with water vapour. Typical applications in this field include:

(i) Chemical and food processing (compressed air supply)
(ii) Air damping plants in textile mills
(iii) Compression of carbonic acid in sugar factories
(iv) Inhalation plant in pharmaceutical processing
(v) Brewery plant
(vi) Gasworks and coke oven plant

The main limitations of the liquid ring pump are:

(i) It is only suitable for pumping gases
(ii) It has a low overall efficiency
(iii) It has a limiting minimum speed below which the liquid ring will collapse and the pump will no longer operate. It is also somewhat limited in maximum speed (rev/min) depending on the size of the pump.

The most widespread application of the liquid ring pump is its incorporation as a priming device in single and multi-stage centrifugal pumps, mounted in parallel with the main pump, or in vacuum duties.

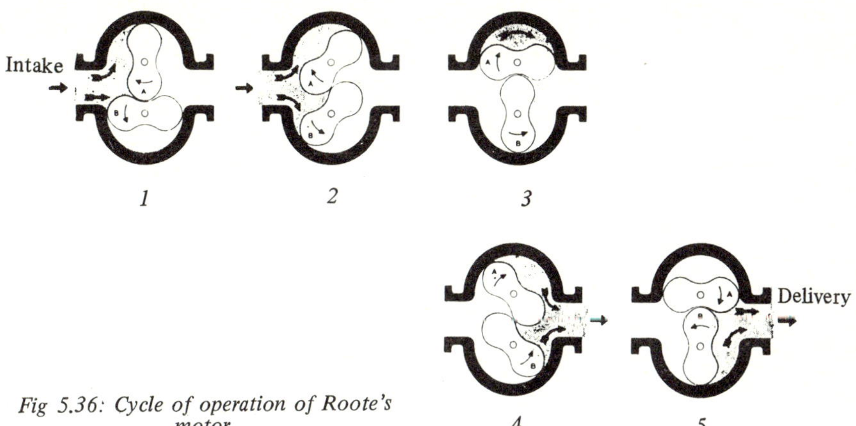

Fig 5.36: Cycle of operation of Roote's motor.

THE ROOTS PUMP

The Roots pump handles only gases, employs two symmetrical figure-of-eight rotors, both driven, but rotating with positive clearance between them and the pump casing. Synchronous rotation of the motors in opposite directions sweeps a positive volume of gas from inlet to discharge side once per revolution – see **Fig 5.36.** Such a pump can be dynamically balanced and is capable of running at high speeds (*eg* 1 500–3 000 rev/min or more) with minimal noise and vibration. Its main uses are as a low pressure gas (or air) compressor or *blower;* and as a vacuum pump. In the latter category the inherent low compression ratio of a Roots pump means that it is usually combined with a backing pump to pump down from atmospheric pressure (or a number of Roots pumps worked in series).

Overload Protection

The pressure difference possible to achieve with continuous operation of a Roots pump is limited by thermal overload. Overload protection can be provided by built-in pressure switches, or an *overload valve* in a by-pass. An alternative method of protection is to drive the Roots pump via a torque converter.

Pumping Speed

The effective pumping speed of a Roots pump can be calculated from:

$$S = S_{th} \cdot \frac{K_m}{Km + \frac{S_{th}}{S_v} - \frac{S_v}{S_{th}}} \quad (m^3/h)$$

The corresponding suction pressure is:

$$P = \frac{S_V \cdot P_V}{S} \quad \text{(mbar)}$$

where:

- S = effective suction speed of the Roots pump (m^3/h)
- S_{th} = theoretical pumping speed of the Roots pump (m^3/h)
- S_V = pumping speed of the backing pump (m^3/h)
- P = inlet pressure (mbar)
- P_V = fore vacuum (outlet pressure) (mbar)
- K_m = maximum compression ratio

For Roots pumps with an overflow valve, this formula is only valid in the pressure range where the overflow valve is closed, *ie* if the difference $P_V - P$ is smaller than the pressure difference for which the overflow valve is set. In the range of pressure where the overflow valve is opened the pumping speed can be calculated thus:

$$S = \frac{S_V (P + p)}{\Delta P} \quad m^3/h)$$

where:

- ΔP = the pressure difference for which the overflow valve is set
- P = partial pressure

Pressure Range

The pressure range at which the Roots pump has its main application is between 30 and 0.001 mbar. However, it can also be used for processes working between 1 000 and 30 mbar. For the backing pumps the rotary vane or rotary plunger pump is, for both technical and economic reasons, usually the best solution. However a liquid ring pump may also be used as a backing pump in particular circumstances.

6. Special Purpose Pumps

THE CHEMICAL INDUSTRY utilises a vast number of pumps for low pressure process duties, the greater majority of which are end suction volute type centrifugal pumps with either open or closed impellers. Pumps of this type are designed to handle a wide variety of fluids ranging in properties from inert to highly corrosive or hazardous and are generally described as *chemical pumps,* or process pumps (although there is a subtle difference between the two types – see *Process Pumps*). For smaller flow requirements, or special duties, a variety of other types of pumps suitable for handling chemical products or chemically active fluids may be employed. These may or may not be classifiable as chemical pumps – *eg* some may be produced specifically for chemical services, but others may be multi-purpose, general purpose or specific pumps capable of handling the fluids concerned with or without changes in material specifications. The description 'chemical pump' is therefore rather a loose one.

The problem of producing a *chemically resistant* pump can be simplified by reducing the number of pump parts in direct contact with the fluid, but for chemical process work the capacity required normally favours the use of a centrifugal pump of more or less conventional design where all the main parts are in contact with the fluid being handled. To construct such pumps throughout in chemically resistant materials can be extremely expensive if the pump is large, and some materials with high chemical resistance can be temperamental in performance.

The centrifugal pump is, however, basically simple in design so that it can be fabricated with comparatively little machining (*ie* utilising castings) in a wide range of corrosion resistant materials which are difficult to machine – including stainless steels, alloy steels, Monel, Hastelloy, Inconel, titanium, glass, ceramic and *plastic materials.* Alternatively, wetted parts can be cast in cheaper materials such as iron, steel or bronze, and lined or coated with rubber, ceramic, plastics, *etc.*

Ultimate choice may best be based on an overall assessment of initial cost and maintenance costs rather than simply material resistance. The position is considerably easier in the case of small pumps where cost and possible lack of mechanical properties on the part of materials with high chemical resistance are not such significant factors. Again, however, the optimum solution is often a matter of overall economics, taking into consideration the reliability of service required and the cost of shut-downs and normal maintenance costs.

Though simplicity in design might seem a very general requirement, it is especially important in the cases of units, such as pumps that perform a vital and often heavy duty. It is generally accepted that, as in the case of all mechanical devices, the simpler the design the greater the reliability, if only for the mere reason that there are fewer things to go wrong. It is obvious, too, that if a unit is simple in design, manufacturing processes such as casting may be easier and the amount of subsequent machining less so that the initial costs are lower. Also, simple design usually goes hand-in-hand with easy and thorough maintenance, as well as facilitating adjustment and replacement of parts. Again, the fewer the number of parts, and the greater the proportion that are identical in a number of models and so interchangeable, the better.

From the point of view of operation, a simple design usually minimises the number of pockets, cavities and crevices in which corrosion and erosion are particularly prone to occur.

In operation, aeration and cavitation of the liquid must be avoided as far as possible to prevent harmful effect on the materials. Cavitation is a factor that is sufficiently serious in the case of pumps handling water, but very much more so when handling chemicals. There are several causes, the result being a lowering of pressure giving rise to vapour formation, followed by the sudden collapse of the vapour bubbles which cause damage to the pump by pitting. Corrosion and erosion can aggravate the damage.

CHEMICAL PUMPS

A satisfactory design of chemical pump must permit easy stripping down for inspection and replacement of components, as necessary. The true cost of a particular (replaceable) component is given by initial cost + life + replacement factor (stripdown cost multiplied by frequency factor) + shut-down factor (actual cost of shut-down time multiplied by frequency factor). If evaluated on this basis the respective merits of a high cost, long life pump and a low cost, short life pump can be determined. A simple cost x life figure does not give a true picture.

Particular attention may also be required in the design of suitable glands and seals, both to maintain satisfactory working temperatures for the seal materials when handling hot fluids and to eliminate chemical attack or corrosion. Mechanical shaft seals are becoming increasingly favoured for chemical pumps, although they do not necessarily provide complete solutions when handling chemically active fluids, or abrasive slurries.

The problem of sealing is further aggravated where the fluid being handled is highly corrosive or toxic and external leakage must be eliminated. Where this cannot be accommodated by straightforward means a glandless pump may have to be employed; or alternatively the pump can be submerged in the liquid. Both shaft driven and canned motor pumps are used for the latter duty.

PROCESS PUMPS (see also **Table I**)

As with chemical pumps, process work normally demands moderate to high flow rates for which the centrifugal pump is generally superior in efficiency to other types as well as being ideally suited to the speeds normally available from electric motor drives. The standard process pump is thus a true general purpose centrifugal pump which is generally available in alternative materials of construction to suit the widest possible range of fluids to be handled. It is also (usually) designed to be suitable for operating over a wider range of liquid temperatures than an ordinary general purpose pump, or a chemical pump.

TABLE I – EXAMPLES OF CENTRIFUGAL PROCESS PUMP CONSTRUCTION MATERIALS

Description	Casing	Impeller	Shaft	Seals
Standard	Close grain cast iron	Close grain cast iron or bronze	High tensile steel	Packed glands or mechanical seals
Alternatives(in various combinations)	Lined*	Coated*	High tensile steel or stainless steel	Packed glands or mechanical seals
	Cast steel	Bronze or gunmetal	Stainless steel	
	Stainless steel	Stainless steel	Stainless steel	
	Gunmetal	Cast steel	High tensile steel	
	Stainless steel	Epoxy	Stainless steel	
Special constructions	Nickel-chrome alloys	Ni-Cr alloy	Stainless steel or monel	Packed glands or mechanical seals
	Hastelloy	Alloy steel	Stainless steel or monel	
	Aluminium	Aluminium	High tensile steel	
	Plastics (see Chapter 10)	Polypropylene (small sizes)	Plastic coated H/T steel	Packed glands Plain bearings Seal ring
		Other plastics (small sizes)	Stainless steel	Plain bearings and ring seal

Linings and Coatings include:
- Hard rubber (on cast iron or steel)
- Glass (on cast iron or steel) for hygienic services
- PVC (on cast iron or steel)
- PTFE (on cast iron or steel)
- Polythene (on cast iron or steel)
- Nylon (on cast iron or steel)

Note that with non-metallic linings the working temperature of the pump is limited by the maximum continuous service temperature of the lining or coating material.

The design concept differs somewhat from that of chemical pumps for whilst resistance to corrosion is a primary requirement construction is generally more rugged, particularly on non-replaceable parts such as casings. This extra thickness provides a generous margin to accommodate erosion-corrosion or wear. Equally, it favours the use of lower cost materials, whereas the true chemical pump constructed in high-cost materials would avoid unnecessary material volume. This distinction is a fine one, and not adhered to by all manufacturers. Thus with the now common use of stainless steel, particularly for smaller pumps, broad compatibility may be achieved in a single pump design which can be equally classified as a 'chemical' or 'process' pump. Some manufacturers do not distinguish between 'chemical' and 'process' pumps in a particular design offered in a range of alternative materials to meet different requirements of chemical resistance.

With horizontal pumps the discharge is normally vertical with a horizontal inlet (although vertical inlet may be employed in some cases). Since the pump may be called upon to accommodate a wide range of duties covering different working temperatures, centreline mounting is often preferred to minimise distortion due to differential expansion effects. Alternatively the pump may be radially mounted to the motor stool in the case of electric motor drivers, or directly on the motor in the case of smaller sizes. A vertically- rather than a horizontally-split casing may also be preferred as simplifying the casing seal and provide ease of disassembly without disturbing the pipeline connections.

Horizontal ring section pump for pressure boosting duties.

Single-vane centrifugal pump for handling cosettes.

Vertical pumps are also employed, particularly where a saving in floor space is important and/or in-line mounting is desirable. A vertical pump also offers a considerable advantage where the pump may have to operate over a wide range of temperatures since thermal expansion and contraction will have little or no effect on alignment. Disassembly can also be simplified. Thus with stool mounting the complete motor, stool and impeller assembly can usually be withdrawn without disturbing the casing and connected pipework.

Multi-stage centrifugal pumps (both horizontal and vertical types) are employed where higher pressures are required. The pressure available from a typical single-stage centrifugal pump may range up to 500–600 lb/in^2 (35–40 bar). Multi-stage pumps are used for pressures above 300 lb/in^2 (20 bar) and so application of single- and multi-stage machines overlap to a certain extent. The majority of process fluid transfer is generally accomplished at relatively low heads and so single-stage machines are the more common.

Gland cooling may be provided as standard, or as an alternative, so that a standard type of process pump can accommodate a wide range of fluid temperatures. Coolant may be bled from the discharge side of the pump, if the liquid and liquid temperature is suitable, or require a separate supply. The process pump should allow for both possibilities. Mechanical seals may also be offered as alternatives to packed glands, the latter normally being a standard choice because of lower initial cost.

Characteristically, a range or series of process pumps by an individual manufacturer will incorporate a high degree of interchangeability, both of major components and materials, so that the basic design can be assembled in a variety of ways to meet the widest possible range of applications.

GLANDLESS PUMPS

The true glandless pump has no dynamic seals or glands acting as a barrier between the fluid being handled and the outside of the pump and other connect-

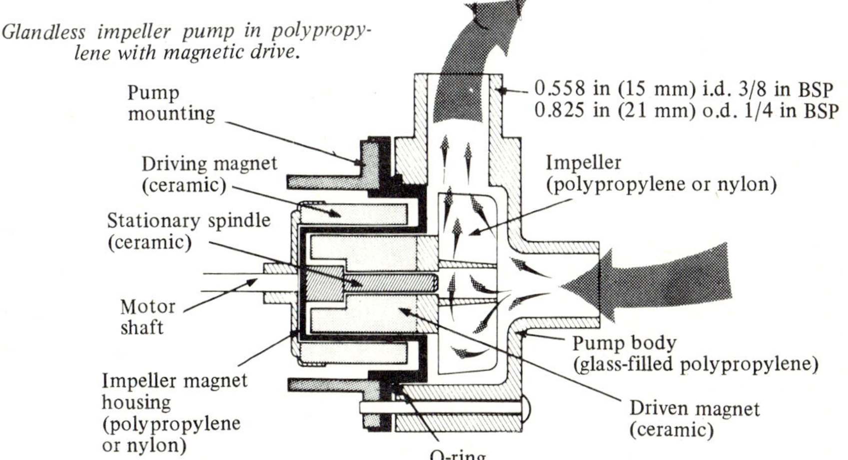

Glandless impeller pump in polypropylene with magnetic drive.

ing parts of the pump. Thus a diaphragm pump is a glandless pump. The description is, however, extended to many other types of pumps which tackle the problem of avoiding gland leakage in other ways, such as:

(i) Submerged, vertical shaft driven centrifugal pumps where no close-fitting shaft gland or seal is required.
(ii) Submersible pumps.
(iii) Canned motor pumps.
(iv) More loosely to various types of constructions where the shaft gland or seal is isolated from the wetted parts of the pump.

BOREHOLE PUMPS

Borehole pumps, or deepwell pumps, are of special design to enable the pump to be lowered to the bottom of the borehole or well until the suction side (or the whole pump) is immersed. The necessary lift to raise the water to surface level is then accomplished on the delivery side of the pump. Basically such pumps are one of two types – (i) shaft-driven, or (ii) submersible – although other types of pumps may be employed for such services.

With shaft-driven pumps the driver is mounted at the surface of the borehole and connected to the pump via a length of vertical shafting. The shaft is normally protected by a sleeve, which can also act as the delivery pipe.

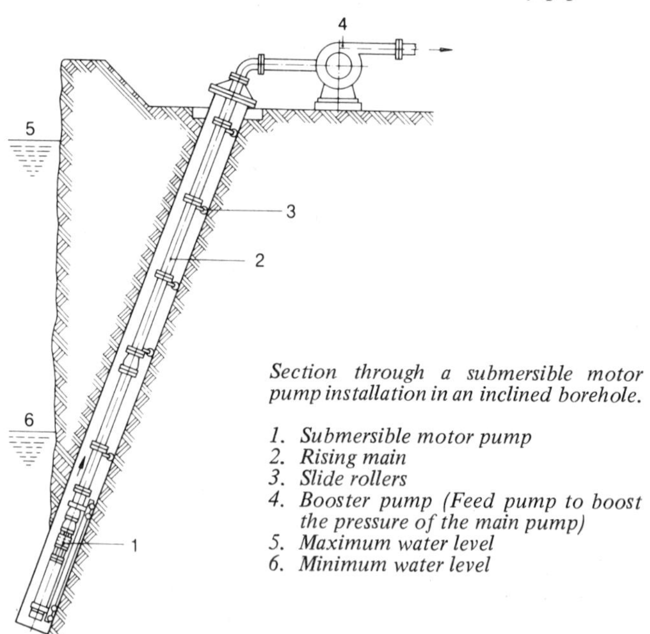

Section through a submersible motor pump installation in an inclined borehole.

1. *Submersible motor pump*
2. *Rising main*
3. *Slide rollers*
4. *Booster pump (Feed pump to boost the pressure of the main pump)*
5. *Maximum water level*
6. *Minimum water level*

Pump, sleeve and driver may be assembled as an integral unit (common with electric motor drives); or driver may be separate and driving the upper end of the shaft through gearing (applicable where the driver must be mounted horizontally, *eg* an i/c engine). Shaft and casing are assembled in sections to make up the total length required.

The shaft is normally supported at intervals along the length of casing, these bearings being of rubber or laminated thermoset plastic when water lubricated – *ie* by the pumped water flowing upwards through the casing. If the liquid being pumped is not suitable as a lubricant then the shaft is further surrounded by an inner tube to isolate it from the liquid flow. This inner tube is then flooded separately with water, or a suitable lubricant, from a separate source.

A centrifugal type pump is the usual choice, although the diameter is necessarily restricted by the size of the borehole available. Economic borehole sizes may range from 6 in (150 mm) diameter upwards, and borehole pumps are produced to match standard borehole sizes. This normally brings the specific speed of the pump within the mixed flow range covered by diagonal pumps where the maximum head available per stage is between 100 and 200 ft (30–60 m). The complete pump is then assembled as a multi-stage unit (usually with axial recuperators) the number of stages being chosen according to the head required. Thus very high head pumps may have up to 20 or more stages, although with improvements in design yielding higher heads per stage, modern borehole pumps tend to have a lower number of stages generating the required head.

It will be appreciated that with the considerable difference in (borehole) size and the number of stages which may be employed, pumps of this type have a wide range of capacities, (*eg* up to 30 000 gal/min (135 m^3/min) or more) and heads up to 1 500 ft (450 m) or more. With unit construction, however, a basic diameter size can be assembled in any number of stages, according to the head required.

This type of pump is not restricted to borehole or deepwell duties. It is equally applicable to shallow wells and also finds applications in other services requiring a high head to be developed by the pump; and for liquids other than water.

Limited use is still made of shaft driven reciprocating pumps for deepwell services. Double-acting pumps of this type are capable of delivery heads of up to about 750 ft (230 m) although they are not generally suitable for boreholes because of the bulk of the pump which has to be accommodated at the bottom of the well.

SUBMERSIBLE PUMPS

The submersible pump is a type which has become increasingly used for a variety of services other than pit draining and water-raising (for which it was originally developed), including the handling of sewage and slurries. Basically it is a centrifugal pump (in multi-stage units for deepwell pumping) with a directly coupled motor, the complete unit being completely immersible. The only connection to the surface is a rising main and a power cable for the motor (or hydraulic lines where powered by a hydraulic motor).

For deepwell services submersible pumps are less costly than shaft-driven pumps, and much more suitable for small sizes if required, as well as being more flexible as regards performance attainable, although capacity range is more restricted.

Submersible pumps are generally classified as *wet motor* or *dry motor* types, the former being the more common. In *wet motor* designs the motor is normally mounted immediately below the pump and the motor compartment filled with oil or clean water when assembled, stator windings being insulated accordingly. The motor, thus operates under controlled and consistent wetted conditions for which it was designed, provided adequate provision is made to exclude outside fluid being sucked in due to thermal expansion and contraction of the casing, or through any other leakage. In *dry motor* designs the motor is mounted above the pump and separated from it by a cylindrical section or air cylinder which has a number of holes in the bottom. Air is trapped in the upper part of this cylinder when the pump is immersed and acts as a barrier to exclude water (or liquid) from entering the motor itself. This, in its simplest form, will only provide temporary protection and in practice the air trapped in the cylinder needs

Sumo submersible pump on test.

periodic charging to replace air losses (*eg* when the water level in the cylinder reaches a critical height). This re-charge is normally accomplished by directing compressed air into the cylinder. On some designs the necessary compressor is located within the pump unit itself. In other cases it is fed from a separate compressor at the surface. Automatic operation can be initiated by a level sensing switch in the air cylinder, although in the former case (built-in compressor) protection is only automatically maintained as long as the pump is running.

Lowering a submersible motor pump into the deep well with the aid of a mobile crane.

The pump itself is essentially similar to that used with shaft-driven units (*ie* usually a diagonal type); but it is far more economic to go on adding stages, if required, to increase the head available. Thus submersible pumps have been produced capable of heads of up to 12 000 ft (3 700 m) for very deep boreholes, and in some cases the number of stages employed have run up to 200–300.

At the other extreme, single-stage submersible units have become a standard type for shallow well pumping, pit drainage, contractor's working, *etc.*

SUBMERSIBLE DRAINAGE PUMPS

This specialised design of pump is closely related to the submersible deepwell pump except that it is intended for low head applications, and comprises, basically, a single-stage centrifugal pump, vertically mounted, with a close coupled electric motor. It originally evolved during World War II for military uses but has been redeveloped during the last thirty years or so as a direct competitor to the typical centrifugal or diaphragm contractor's pump sets powered by diesel or petrol engines. Compared with the latter types it has the advantage of eliminating the necessity of a suction hose, as well as being generally more compact and lighter in weight for a given capacity. Capacities available with portable units range up to about 300 gal/min (70 lit/min) or more with weights of up to 100 lb (45 kg). Larger units are also produced and may be regarded as semi-portable. Still larger units may be used for permanent or semi-permanent installations, lowered in position and raised for inspection, *etc,* by hoisting gear as necessary.

This type is becoming increasingly popular for a wide variety of drainage, de-watering and similar services where a mains electricity supply is available, and is particularly suitable for continuous operation without supervision under automatic control. Overall efficiencies of up to 50% are achieved with the majority of larger units and are thus normally appreciably higher than those achieved with engine-driven pump sets of similar capacity.

BOILER FEED PUMPS (see also **Table II**)

Boiler feed represents a specialised service, associated with steam power plant, the pumps used invariably being specially designed for this application, as boiler feed-pumps. The pump *type* employed is largely influenced by the plant (*eg* low-pressure, medium-pressure or high-pressure) and the capacity required, although there is an overlapping range up to about 400 gal/min where both reciprocating and centrifugal pumps may be employed. In the main, however, centrifugal volute, centrifugal diffuser and centrifugal regenerative types cover the bulk of capacity requirements. Pressure requirements are then met by multi-stage machines, with up to 12 stages quite common. Actual capacity is usually based on the maximum allowable overload rating of the boiler, plus a generous safety factor (commonly 20%). Pressure must equal boiler pressure plus friction losses, with an additional allowance for supplying the boiler with the safety valve open. Typically the pressure is of the order of 20 to 25% higher than the rated boiler

Boiler feed pump and turbine on common baseplate. (Sulzer)

pressure, and in the case of large plant requirements may approach 6 000 lb/in^2 (720 bar). A further factor which aggravates constructional problems is that the water temperature may range up to 700–750°F (370–400°C).

TABLE II – SELECTION RANGES FOR BOILER FEED PUMPS

Pump Type		Capacity gal/min (m^3/hour)					Pressure Range lb/in^2 (bar)			
		up to 200 (up to)	200–400	400–600	600–1000	over 1000	up to 410 (up to 28)	400–1200 (28–84)	1200–2000 (84–140)	2000–6000 (140–420)
Reciprocating		√	√				√*	√*	√*	√*
Regenerative:	Single stage	√					√			
	Multi-stage									
Centrifugal:	Volute	√					√			
	Multi-stage		√	√	√	√	√	√	√	√
Centrifugal:	Diffuser		√	√	√	√	√	√		
	Multi-stage			√	√	√		√	√	
Barrel-type:	Multi-stage								√	√

*Multi-cylinder

Multi-stage centrifugal type boiler-feed pumps are commonly classified by construction, *viz* –

(i) *Split casings* – which in general are restricted to a maximum pressure of about 1 500 lb/in^2 (105 bar) or less on account of the difficulty of sealing the casing joint, although some higher pressure units are made with this form of construction. Pumps of this type are widely favoured in America.

(ii) *Ring or unit construction* – where each stage comprises a separate casing with its suction chamber and delivery chamber. A multi-stage casing is then built up by bolting together the required number of units with external tie bolts. Gasket type seals may be used between the casings for pressures up to about 1 000 lb/in^2 (70 bar). For higher pressures the jointing surfaces are usually ground and lapped. Whilst this form of construction is straightforward, a difficulty is the possibility of distortion of the complete pump body under pressure and temperature changes. Thermal conditions may call for preheating or a gradual warm-up of the pump on certain designs. On high pressure designs, some pressure relief may be obtained by dividing the length of the assembly into high-pressure and low-pressure sections by means of an intervening partition. Such pumps are capable of working up to the highest feed pressures required without aggravating distortion problems.

Unit type construction is particularly favoured by European pump manufacturers.

(iii) *Barrel-type casings* – here an inner casing fits into an outer barrel, pressure acting on the inner casing then providing the force necessary to seal joints between individual stages. This pressure is equal to the pressure in the delivery branch, but is partly balanced by the internal pressure in the stages so that thick inner casing walls are not necessary. A further advantage is the small number of high pressure fluid joints which have to be sealed and so this form of construction is particularly suitable for high pressure designs. Barrel-type pumps are also less liable to distortion, although they are generally heavier and more expensive to produce. They are a particularly popular type in America for medium and high pressure services, but less so in European productions.

The most common type is the horizontally split casing with suction and delivery branches located in the bottom half of the casing. The upper half can then be removed for servicing without disturbing the pipework. Vertically mounted split casing pumps are, however, coming into greater use, particularly where saving in floor space is important.

Constructional Materials

Cast iron is the chief material employed for casings for pressures up to about 1 000 lb/in^2 (70 bar) and water temperatures up to 250°F (120°C). For higher temperatures SG cast iron would normally be preferred. Cast steel may be used for casings for higher pressures, although this material is not particularly resistant to corrosion by feed water. Thus 5% chromium steel is a more common choice, or 13–15% chrome steel.

Impellers (and diffuser rings in diffuser type pumps) are commonly made of bronze for lower pressures and temperatures; and 13–15% chrome cast steel for higher pressures and temperatures. Wear rings, *etc,* are normally made of alloy steel or chrome-iron. Shafts are invariably of high tensile steel.

Shaft seals present a particular problem, largely because of the change in working temperature which may be involved. Conventional glands and seals may not be satisfactory and expansion glands may be called for to cope with high pressures and temperatures. Mechanical seals are generally to be preferred and are coming more into favour, although special gland designs continue to be used.

Number of Stages

To some extent the number of stages employed in a particular design must represent a compromise between cost and reliability (favouring a small number of stages), and the pressure which can be realized with the type of pump concerned. The latter is largely governed by speed, particularly when the pump is driven by an electric motor limiting synchronous speed to 3 000 rev/min (50 cycle supply) or 3 600 rev/min (60 cycle supply). In such cases about 300 lb/in^2 (21 bar) per stage is a maximum, although higher speeds (and higher pressures per stage) can be realised by geared-up drives. In the latter case it may still be desirable to limit speed to about 5 000 rev/min maximum for mechanical reliability. Higher speeds are possible with steam turbine drives.

Another method of reducing the number of stages required is to increase the impeller diameter, but this almost inevitably results in loss of efficiency and so is seldom employed. A more practical approach where it is desired to reduce the number of stages is to employ a high pressure main pump and a low pressure booster pump – *eg* see **Fig 6.1**. Both may be driven by the same driver, although it may be advantageous to run the booster pump at a lower speed powered by its own driver, although this increases the pressure and working temperature of the main pump.

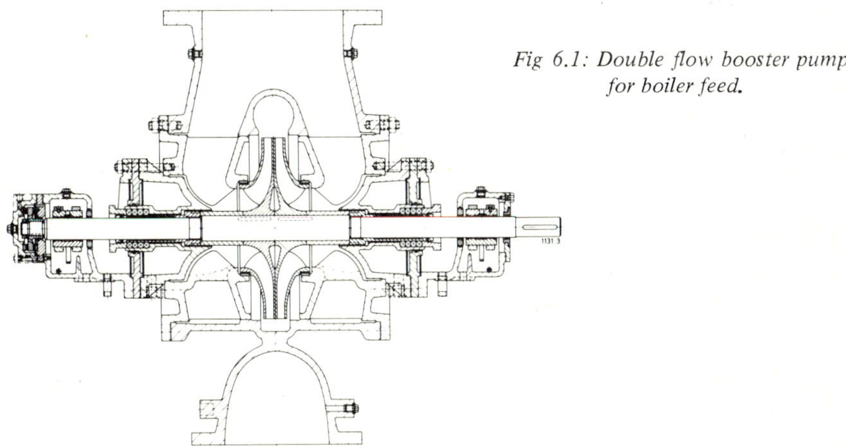

Fig 6.1: Double flow booster pump for boiler feed.

Capacity Requirements

General practice favours the use of one pump per boiler, capable of meeting the capacity requirements of that boiler. Where the demand fluctuates, however, it may be better to employ two or more feed pumps in parallel. Three half-capacity pumps in parallel would always provide one pump available for stand-by.

Characteristics required of a boiler feed pump are a stable and reasonably flat H-Q curve with high efficiency at the working point (80% efficiency being the usual maximum attainable on larger sizes). Efficiency is very important as this will affect the heat balance of the system.

Condensate Pumps

Condensate pumps are almost invariably of centrifugal type, although reciprocating pumps were originally also used for such services. The pumps are of special design to prevent air intrusion and invariably work with flooded suction, *ie* mounted below the lowest water level in the condenser, to minimise the chances of cavitation. Vertical pumps offer an advantage in this respect since they increase the available positive suction head, as well as reduce floor space occupied by the pump; but horizontal pumps remain the more usual choice.

The pressure in the condenser is a low order of vacuum and the temperature of the condensate is close to that of the saturated steam for a given pressure hence the positive head available must at least be equal to the NPSH required by the pump, if cavitation difficulties are not to arise. The pump is thus sited accordingly. Two typical systems are shown in **Figs 6.2** and **6.3.**

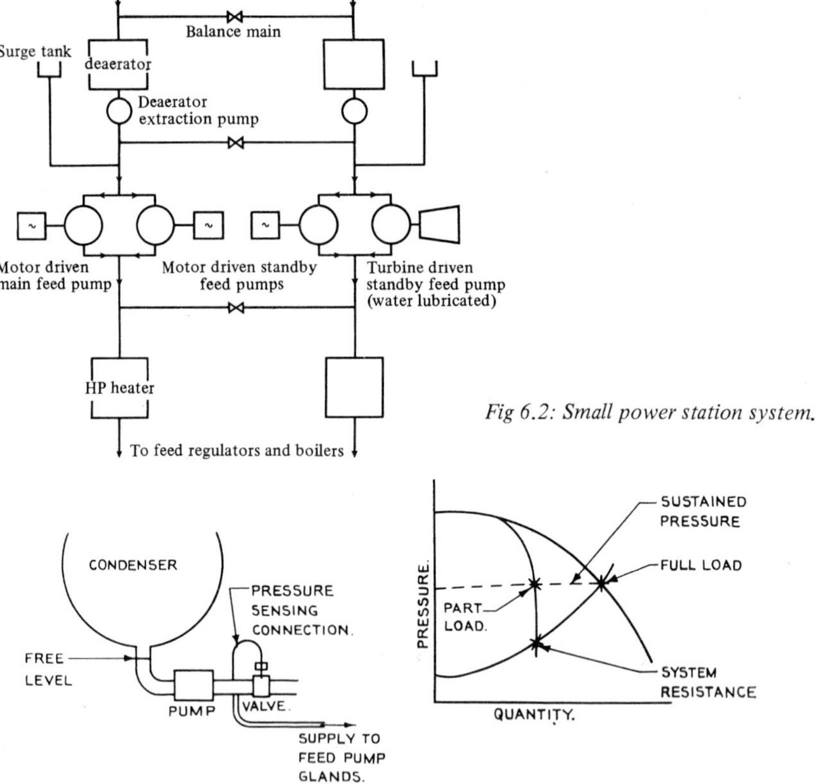

Fig 6.2: Small power station system.

Fig 6.3: Modern power station practice providing a supply of water under pressure to seal the feed pump glands.

TABLE III – TYPES OF CONDENSATE PUMPS

Duty	Pump Type(s)	Remarks
Low pressure	(i) Single stage centrifugal, horizontal (ii) Single stage regenerative	End suction, oversized
Low to medium pressure	(i) Two-stage centrifugal, horizontal (ii) Two-stage centrifugal, vertical (iii) Multi-stage regenerative	End suction, oversized
High pressure	(i) Multi-stage centrifugal, horizontal (ii) Multi-stage centrifugal, vertical	Middle section is commonly first stage May be similar to borehole pumps

Pump types most commonly employed are summarised in **Table III**. In the case of multi-stage machines, and particularly with horizontal pumps, the suction of the first stage is designed for a higher discharge than the nominal requirements, either using a different impeller diameter or a double-entry impeller (or both). This has the advantage of making the pump less susceptible to cavitation troubles, *ie* reduces the NPSH required by the pump, although it does lower the overall efficiency. Common design features include:

(i) Water-sealed glands or stuffing boxes to eliminate air intrusion.

(ii) Vented suction chambers to prevent vapour binding.

(iii) External sealing of stuffing boxes to prevent air intrusion when the pump is idle.

Electric motor drive is virtually standard for all sizes of condensate pumps, except for marine installations where turbine drives may be employed.

Condensate-booster and hotwell pumps follow, basically, the same requirements.

METERING AND PROPORTIONING PUMPS (see also Tables IV and V)

Metering or proportioning duties are normally accomplished with special designs of piston-diaphragm pump heads where high accuracy is required. Plunger-type diaphragms may be used where higher pressures are required; or a bellows pumphead where greater capacity is required from a given size of pump (see **Fig 6.4**). Other positive displacement pumps are not excluded for such duties, *eg* rotary pumps may be used for metering viscous fluids, but cannot reproduce the accuracy of a true metering pump which is normally of the order of plus or minus 1% or better.

In the main, metering pumps are small, with capacities in the range 0.5–440 gal/hr (100 ml/hr–2 000 lit/hr). Very much smaller pumps are also produced, known as *micro-metering pumps;* however larger sizes up to 15 500 gal/hr (70 000 lit/hr) for large scale proportioning duties. Since metering and proportioning duties commonly call for different volumes to be delivered at different times, most metering and proportioning pumps are of variable-capacity type.

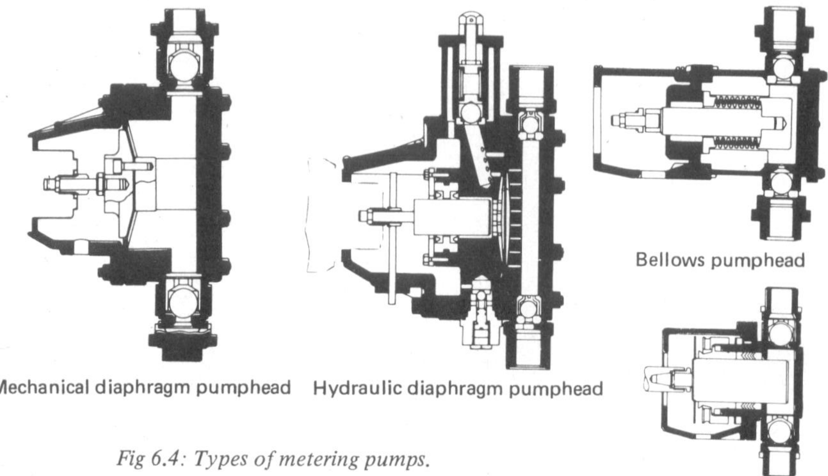

Fig 6.4: Types of metering pumps.

TABLE IV – PLUNGER TYPE METERING PUMP MATERIALS

Pump Component	Fluid Handled		
	Acid	Alkaline	Water, oils and general
Pump block	Ebonite- or polythene lined** Cast iron or steel	Cast iron	Stainless steel*
Piston	Porcelain or glass	Stainless steel	Stainless steel*
Gland or packing	Polythene or synthetic rubber	Synthetic rubber or PTFE	Synthetic rub-rubber
Valve housing	Ebonite-lined. PVC or polythene	Mild steel	Stainless steel *
Valves	Glass	Steel	Stainless steel*
Valve seats	Ebonite	Stainless steel	Stainless steel*
Suction connections	Ebonite or PVC	Mild steel	Stainless steel
Delivery connections	Ebonite or PVC	Mild steel	Stainless steel

* Other materials include Hastelloy, Monel, Titanium, Carpenter 20 Alloy, *etc*, according to specific duty.

** Other linings for cast iron include Rubber, Hypalon, Penton, Polypropylene, PTFE, Nylon and Glass.

There is also a distinction between a true metering pump and a proportioning pump. A metering pump can provide both duties, but a true proportioning pump has a delivery which is made directly proportional to one parameter (usually flow rate).

Piston or Plunger Metering Pumps

Piston-type metering pumps are of straightforward plunger design with variable delivery characteristics achieved either by adjustment of stroke or speed, or both. Double suction and delivery valves are commonly employed since this safeguards constant delivery in the event of one valve not closing properly due to some contaminant in the fluid lodging on the valve seat. The sealing problem can be reduced to that of the plunger seal, which is critical if accurate metering performance is to be maintained. Power is normally from an electric motor, driving through reduction gearing as necessary. A single motor may drive a number of

TABLE V – COMPARISON OF METERING PUMP TYPES

Pump Type	Advantage(s)	Disadvantage(s)
PLUNGER	High accuracy with good repeatability. Capable of delivering high pressures. High reliability.	Susceptible to leakage, or corrosion handling corrosive fluids.
MECHANICAL-DIAPHRAGM	Least expensive. Can handle most fluids. Glandless.	Accuracy not so good. Diaphragm subject to fatigue failure. Limited to low pressure deliveries.
PISTON-DIAPHRAGM	High accuracy. Good diaphragm life. Can handle most fluids. Glandless. Readily rendered in double-diaphragm form for fail-safe characteristics.	Diaphragm may be subject to wear handling fluids, including particulate matter.
ELECTRO-MAGNETIC DRIVEN DIAPHRAGM	Particularly suitable for micro-metering with precise pulse operation. Glandless.	Limited capacity. More complex control system, (needs digital signal input)
BELLOWS	Offers greater capacity for a given physical size of pump head. Suitable for higher temperatures. (Bellows can be metal or elastomer). Glandless.	
PERISTALTIC	Simple construction. Glandless. Can handle most fluids.	Accuracy can be affected by temperature changes, (*eg* expanding or contracting the plastic tube).

individual metering pumps assembled in a block. In some designs the plunger is hydraulically operated rather than mechanically driven, this method also simplifying stroke control.

Displacement is directly proportional to the stroke. Linear stroke (and thus displacement) control can be provided by a micrometer screw altering the throw of an eccentric, or some similar means. The main requirement is that any such method of adjustment should eliminate lost motion as far as practicable, and preferably also provide true linear adjustment (although this is not absolutely necessary). Stroke control can be extended over the full range, but at very small values of stroke will tend to become unreliable and critical. The practical working range thus normally ceases a little above zero stroke, *ie* at a positive value of delivery rather than zero delivery. Adjustment for any capacity over the working range becomes less critical with increasing capacity; and is also the lss critical the higher the stroke: bore ratio. For this latter reason long stroke small bore plungers are normally adopted for metering pumps.

Piston-Diaphragm Metering Pumps

Diaphragm heads in the case of piston-diaphragm pumps may incorporate single or double diaphragms. In the latter case the space between the two diaphragms is filled with an inert fluid, giving the unit 'fail safe' characteristics. That is, in the event of failure of one diaphragm the fluid being handled cannot become contaminated with oil from the hydraulic circuit. The use of double diaphragms with an inert or buffer fluid between them also provides a simple means of displacement control, *ie* increasing or reducing the buffer fluid volume will decrease or increase, respectively, the stroke of the diaphragm. Alternatively, variable displacement can be achieved by varying the stroke of the piston pump in the hydraulic circuit.

Again, a number of metering heads can be mounted in a common block actuated by a single pump. In this case the pump only needs a single hydraulic line to connect to the head and can be remote from the pump block, if necessary.

Where the capacity range required is not attainable by basic adjustment it can be extended by varying the speed of the pump. A typical stroking speed for metering pumps is of the order of 100 strokes per minute, but this is capable of being varied over a wide range provided steady speeds can be held. When the metering pump is handling more viscous fluids, speed reduction is usual, a typical working speed in this case being 50 strokes per minute, or less.

Micro-metering pumps may exploit speed control more completely to achieve deliveries which may only be of the order of a few cc per hour. To avoid having to use an excessively small size of pump, impulse timing may be applied to operate the pump periodically instead of continuously to deliver the required quantity in the stated time.

All true metering pumps are precision units and must be handled and maintained accordingly. They should be selected with regard to the specific duty required, although many may be multi-purpose units as regards the types of fluids they can handle and the deliveries available. It should be noted that unless the design specifically claims otherwise, no metering pump will provide positive calibration and positive output unless working against a positive head.

Complete Units

It is becoming increasingly common to supply fully packaged metering or dosing units consisting of pump(s), storage tank(s), pipework, controls and instrumentation.

Control may be pneumatic, electronic, or a combination of both, choice normally being based on plant conditions and requirements.

Other Special Purpose Pumps

The following Table gives a condensed reference to other pump types commonly described by purpose or application. In many cases further information on the type(s) of pumps involved is given elsewhere – check against the pump type(s) mentioned.

TABLE VI

Description or Service	Pump Type(s)	Remarks
ALKALI PUMPS	Chemical – generally suitable Process – may be suitable Specific types: (i) centrifugal (ii) reciprocating (iii) rotary (iv) diaphragm	All-iron construction for general applications. Lined pumps may be employed for more corrosive alkalis – types (i) and (ii), but not ceramic linings. Acid pumps are often equally suited for handling alkalis. Plastic pumps – types (i) and (iii) for smaller capacities.
ASH PUMPS	Single-stage centrifugal special designs – rubber lined	(i) Open suction entry and special impeller (ii) Water-sealed glands usual.
BRINE PUMP	All-bronze or bronze-fitted general purpose pumps (all types)	Suitable for sodium chloride brines.
	All-iron general purpose pumps (all types)	Suitable for calcium chloride brines (except at low temperatures)
	Process pumps (all types)	For handling foodstuffs (but check materials for possible contamination of product).
	Flexible vane pumps	Simplest type for low capacity requirements.
	Diaphragm pumps	Particularly for duties where abrasive solid contaminants are present

cont...

TABLE VI (contd.)

Description of Service	Pump Type(s)	Remarks
CARGO PUMPS	(i) Reciprocating (steam, Duplex) (ii) Single-stage centrifugal (iii) Two-stage centrifugal	For crude and bulk oil cargo – commonly used in association with additional stripping pumps. Less widely used than (ii).
CIRCULATORS (Accessories)	(i) Centrifugal (ii) Axial flow (small size)	Vertical type preferred for in-line mounting. Require smaller motor than centrifugal pump and generate lower water velocities for the same driven speed.
CONCRETE PUMPS	Reciprocating – special design	Low speed piston pumps with hopper feed. Pipeline bore and pump delivery branch 3 to 4 times size of aggregate
COOLANT PUMPS	(i) Centrifugal (ii) Gear (iii) Flexible vane	Special impeller design usual to suit fluid being handled and nature of contaminants likely to be present. Suitable only for clean fluids with lubricating properties. May be used with dirty oil fluids if the pump is protected by screening or filtering (*eg* with magnetic filters). Suitable for all fluids and can handle a certain amount of solid contaminants.
DAIRY PUMPS	Chemical pumps and generally suitable Process pumps – limited suitability Specific types: (i) centrifugal (ii) Reciprocating (iii) Rotary	Materials selected to give sterile handling. Pump designed to eliminate dirt traps, *etc*; also for ease of dismantling and cleaning. Moderate to high deliveries of clean fluids. Alternative to centrifugal. Generally preferred for more viscous fluids such as cream, cheese, condensed milk, *etc*.
DREDGING PUMPS	Single-stage centrifugal (usual)	Armoured construction and/or replaceable wear parts. Impeller design and casing clearances selected according to the nature of the ground.

cont...

TABLE VI (contd.)

Description of Service	Pump Type(s)	Remarks
FIRE PUMPS	Portable pump sets	May be specifically designed as fire pumps or general purpose pumps.
	(i) Centrifugal pumps	Flat H-Q curve desirable for constant pressure at different flow rates. Steep H-Q curve for high pressure at low flow rates. Suction head should not exceed 15 ft (5 m). Max. pressure with centrifugal pumps usually 200 lb/in^2 (30 bar).
	(ii) Rotary pumps	Pressures up to 100 lb/in^2 (70 bar). Suction head should not exceed 20 ft (6 m).
	(iii) Reciprocating pumps	Less used, but particularly suitable for high pressures and higher suction lifts.
FOODSTUFFS	(i) Centrifugal	'Bladeless' impellers for handling delicate solids in suspension. Special single- or two-blade impellers for handling other solids in suspension. Standard impellers for handling clean non-viscous liquids.
	(ii) Reciprocating	Suitable for handling a wide range of viscosities: also suitable for handling chemically active fluids.
	(iii) Rotary (gear)	Oils, soap solutions and other liquids having lubricating properties.
	(iv) Rotary (other types)	All types of clean liquids, particularly more viscous fluids.
	(v) Lobe rotor	Handling sensitive fluids.
	(vi) Screw pumps (rigid screw and eccentric screw)	Wide application for handling all types of foodstuffs. Sterile applications; particularly easy to clean.
	(vii) Diaphragm	
	(viii) Peristaltic	Sterile applications; handling delicate fluids; but low capacities
	(ix) Process or general purpose pumps	Pumps of this class with stainless steel construction are widely applicable, provided they are easy to clean.
GRAVEL PUMPS	(i) Centrifugal – chokeless impeller (usual)	Hard-material construction essential. Replaceable wear parts fitted. Impeller design and clearances must be selected with regard to the size of the solids to be handled.
	(ii) Diaphragm pumps	Contractor's type pumps.

cont...

TABLE VI (contd.)

Description of Service	Pump Type(s)	Remarks
LIQUID GAS PUMPS	(i) Reciprocating (ii) Diaphragm	Special designs of piston pumps with refrigeration. Double-diaphragm type with refrigeration and stainless steel diaphragms. *Note:* special attention must be given to abrasion or clogging problems which may arise from crystallisation (solidification) of the liquid gas (*eg* carbon dioxide).
MOLASSES PUMP	Suitable types: (i) Internal gear (ii) Screw (iii) Lobe rotor	 Number of teeth selected according to viscosity of molasses. Very suitable and adaptable. With suitable rotor design. *Note:* all molasses pumps are generally operated at reduced speed with increasing viscosity calling for reduction gearing and/or variable speed control. Ease of dissassembly for cleaning is a desirable design feature.
MOLTEN METAL PUMPS	(i) Mechanical – special design Electromagnetic – special designs: (i) a.c. conduction (ii) d.c. conduction (iii) Helical induction (iv) Linear induction	Impeller type with glandless construction and provision to maintain melt temperature. For heavier metals. For high resistivity metals. For low resistivity metals (liquid metal isolated) For low resistivity metals (liquid metal isolated) *Note:* all are essentially specialised designs.
MUD PUMPS	(i) Contractor's pumps (ii) Heavy-duty vane pumps (iii) Hydraulically-operated reciprocating pump with cutting cylinder	For general applications. For handling very thick muds and semi-solids. True 'mud pump' for handling very stiff solids.
MUCK PUMPS	(i) Contractor's pumps (ii) Centrifugal with non-choke/impellers (iii) Diaphragm	For general applications. Lined or armoured. Limited capacity.
PIT DRAINAGE	(i) Immersible pumps (centrifugal) (ii) Portable pump sets	May be fitted with automatic stop-start controlled by water level. For temporary duties.

cont...

TABLE VI (contd.)

Description of Service	Pump Type(s)	Remarks
ROUSING	(i) Rotary (ii) Screw (iii) Centrifugal	May be portable unit. Open impeller type: may be portable unit.
SAND PUMP	(i) Reciprocating (ii) Centrifugal – chokeless impeller	Little used. Main type used, with lined or armoured casings and replaceable wear parts. Water-sealed glands or mechanical shaft seals.
SLIP PUMPS (Ceramic slip)	(i) Single screw pumps (ii) Screw and platewheel pump (iii) Rotary	Elastomeric stator. Less liable to clog than plain screw. Limited application because of high wear. *Note:* this duty specifically calls for the use of a positive displacement pump.
STRIPPING	(i) Reciprocating – steam driven (ii) Single-stage centrifugal	Usually 'Duplex' type. Vertical, double-acting.
STUFF PUMP	Centrifugal (usually)	Centrifugal pump with impeller designed to handle paper pulp, *etc*.
SUMP PUMPS	(i) Immersible (centrifugal) (ii) Submersible (centrifugal) (iii) Portable pump sets	Vertical type with extended shaft and electric motor drive. May readily be adapted to automatic operation with a pressure switch. Close-coupled with electric motor drive. For temporary duties. *Note:* exactly the same types of pumps may be specified as 'Cellar Pumps' and 'Sump Pumps' the duties being the same except that sump pumps may range to higher capacities.
SOLIDS PUMPS	(i) Centrifugal – chokeless impellers (ii) Heavy-duty vane (iii) Diaphragm (iv) Peristaltic	See Chapter 9 for more specific information as to suitability based on nature of solids *etc.* Suitable for handling semi-solids at low speeds. Suitable for corrosive media. Suitable for corrosive media and small capacities.
RESIN PUMPS	(i) Reciprocating, at slow speeds (ii) Rotary (iii) Diaphragm	Limited application. Suitable type must be selected with regard to the nature of the fluid. Screw-type or lobe rotor generally preferred for higher viscosities and/or sensitive fluids. Limited working pressure. *Note:* resins are generally classified as viscous fluids – see Chapter 9.

7. Hand Pumps

THE ORIGINAL TYPE of hand pump and shallow well pump developed for water raising was a reciprocating piston pump (force pump), operated by a lever action. The type still survives, but in less massive constructions and with more attention to cylinder bore finishing and piston sealing (normally employing cup seals).

The overall efficiency of such a pump in good condition is of the order of 50 to 60%, delivery rate being related directly to cylinder bore and stroke and the maximum rate of stroking. The larger the displacement of the cylinder the longer the handle became and, generally, the greater the stroke:bore ratio. Such pumps tended to become cumbersome and tiring to work, although they were relatively simple and inexpensive to construct. The type survives in single cylinder designs described traditionally as lift pumps, lift-and-force pumps when fitted with an air chamber and draw-off cock; and in parallel cylinder designs (Duplex) operated by a rocking motion of the hand lever.

Simple Plunger Pumps

Simple plunger pumps with the handle attached directly to the piston rod are produced for a variety of light duties (*eg* bilge pumping on small pleasure craft). These are of 'bicycle pump' proportions with an outlet at the upper end of the barrel and a very high stroke:bore ratio. Modern pumps of this type are usually all-plastic construction, including the piston rod. Efficiency can be quite high, but delivery is relatively low, limited by practical sizes for such pumps and the rate of stroking that can be achieved with a directly operated (manual) reciprocating motion.

Semi-Rotary Hand Pumps

The semi-rotary or wing pump as it is commonly called is a standard form of handpump for water services (*eg* hand operated bilge pumps) and similar duties. The basic design is shown in **Fig 7.1**. The casing is circular and encloses a wing cast integral with or mounted on the pump spindle. Each side of the wing is fitted with a simple flap valve. The bottom of the casing forms a V-shaped chamber or suction box, each side of which is fitted with a flap valve.

The wing is given an oscillatory movement via the handle. Movement in one direction produces a depression in one chamber sucking in fluid via the suction

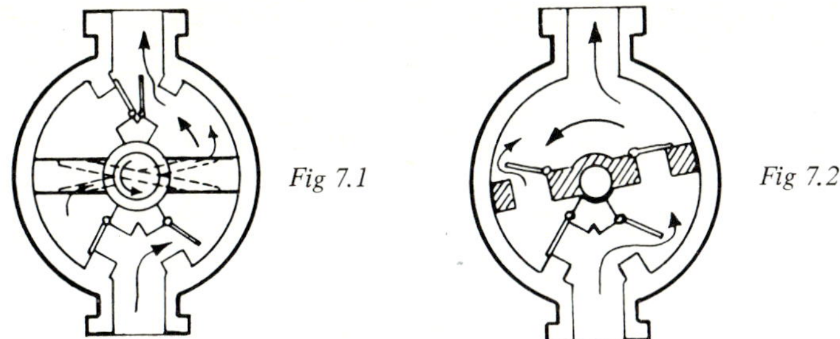

Fig 7.1

Fig 7.2

box and flap valve on that side and compression in the other chamber to pass fluid already in the pump into the upper part of the casing via the flap valve in the wing. Movement in the opposite direction reproduces the operation in reverse, and at the same time forces liquid in the pump chamber up through the outlet. The pump is thus double-acting.

The capacity (per cycle) can be approximately doubled by adopting quadruple action. Here the casing is modified to incorporate an additional valve box or outlet divider which retains liquid in the delivery pipe and has flap valves opening inwards into the main chamber. The wing in this case does not have flap valves and flow through the wing is via suitably shaped ports – **Fig 7.2**. The effort required is increased in the same proportion as the capacity and this may be offset by the use of a longer handle (which will also automatically tend to reduce the rate of stroking).

Diaphragm Hand Pumps

Diaphragm pumps are generally better suited than wing pumps for achieving higher capacities as handpumps. They also have the advantage that they can handle dirty liquids with less chance of blockage or damage and much reduced wear handling abrasive solids in suspension. Using modern elastomeric materials, diaphragm life is long and such pumps are generally reliable in service. They are designed for simple disassembly to clear blockage, or replace internal parts if necessary.

To achieve maximum stroking rate with minimum effort the handle on a diaphragm pump is usually long. About 40 double strokes per minute is a practical maximum for extended periods of working, or rather less on larger pump sizes. Actual sizes may range from about 2 in (5 mm) diameter up to about 10 in (250 mm). Typical delivery rates obtainable are of the order of 15–20 gal/min (3.5–5 lit/min) per inch (25 mm) diameter; or double this with twin diaphragms.

Rotary Hand Pumps

Instead of lever action, some handpumps may employ a fully rotary or 'grinder' action, the input lever being rotated continuously. The required oscillating or reciprocating motion to drive the pump element is then derived by internal mechanical linkage, or the pump itself may be designed as a rotary piston type. Rotary-action handpumps may be designed as high capacity units but, in general, this action is more tiring than handlever action for generating a similar output continuously. Some handpumps are also produced with true rotary pumping action with an extended vertical shaft and cranked operating handle. Both capacity and efficiency is very low unless the input shaft is geared up to the pump spindle, which complicates the design. They are little used compared with the other standard handpump types.

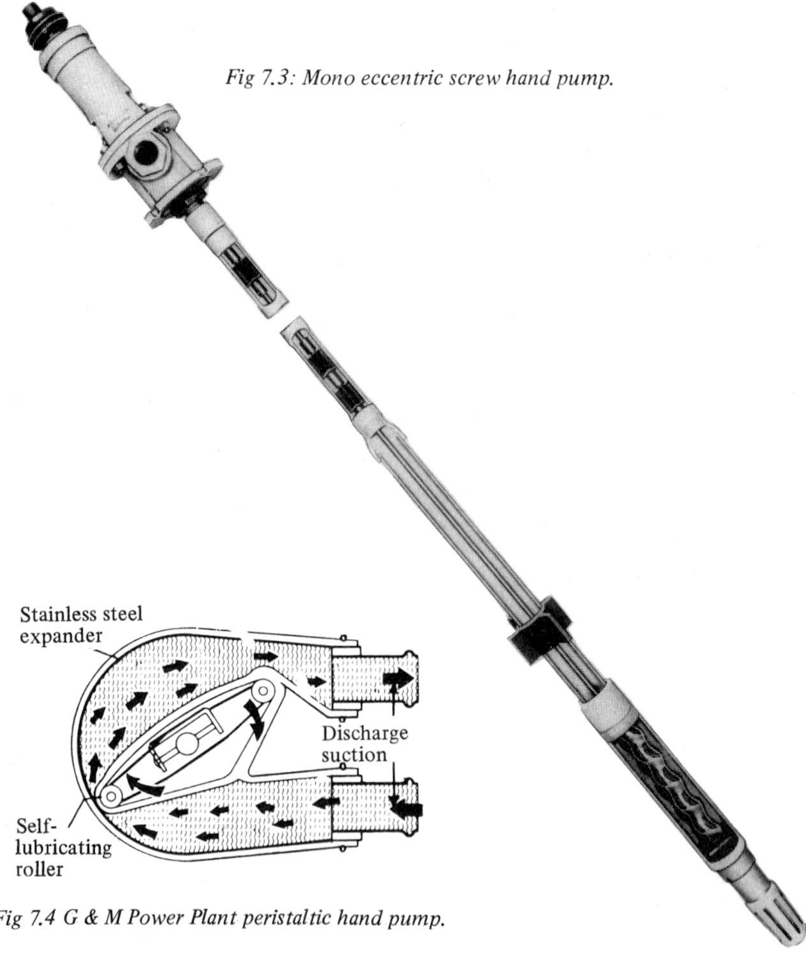

Fig 7.3: Mono eccentric screw hand pump.

Fig 7.4 G & M Power Plant peristaltic hand pump.

The main reason for not adopting a true rotary pump as a hand pump is that a geared-up drive is necessary in order to achieve reasonable pump speed, making the cranking action more difficult and introducing mechanical power losses. A notable exception is the elastomeric stator screw pump which, because of the length of rotor which can be used, can give reasonable delivery rates at low rotational speeds – *eg* see **Fig 7.3**.

Peristaltic Hand Pump

An example of a peristaltic hand pump is shown in **Fig 7.4**. Although the rate of working of such a pump is limited it has the advantage that it can be used to pump corrosive fluids since the product handled is contained entirely within a plastic tube and no other parts of the pump are wetted.

Hydraulic Hand Pumps

Handpumps for generating hydraulic pressure in closed oil circuits are (almost) invariable of reciprocating type. These have a particular application for providing hydraulic power for lifting duties, for portable hydraulic equipment and other hydraulic systems designed for manual operation. Handpumps may also be used as a source of test pressure for hydraulic circuits; or as a source of emergency power in the event of failure of the main pump.

The piston size of such high pressure pumps is related to the pressure required and the mechanical leverage, consistent with a suitable hand load. A typical lever movement is shown in **Fig 7.5**, length, and travel or angular movement of these proportions being suitable for continuous manual operation at horsepower inputs up to the order of 0.05.

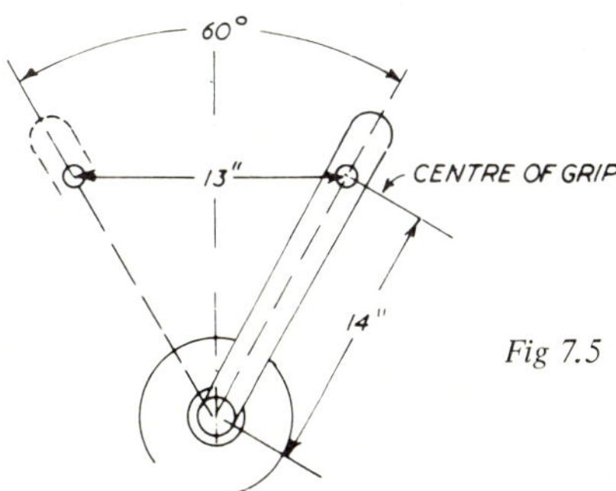

Fig 7.5

Either single- or double-acting piston pumps may be employed, the latter providing double the delivery per complete cycle but demanding the working against pressure load in each direction. The latter condition can be alleviated by making the pump a two-speed or two-stage design to give high delivery at light loads, the pump output being reduced at a predetermined pressure level to reduce the load required to complete pumping up to the final high pressure required. Two basic methods of achieving this are:

(i) A mechanical system to alter the leverage and the stroke of the pump at a suitable changeover point.

(ii) The use of two separate pistons and cylinders, one providing high output up to a moderate pressure when it is unloaded and pumping up is completed by the second smaller unit at reduced output.

Output and Power

The output of a handpump depends primarily on the power capacity of its human operator. The average adult human can develop something like 1/10 horsepower working continuously for long periods or up to 1/4 horsepower in short bursts. Equivalent hydraulic horsepower likely to be developed is, therefore, of the order of 0.05 for continuous working of a typical handpump.

In the case of a lever operated pump maximum hand load usually arrived at is 60 lbf (30 kgf), this being found generally acceptable for continual working without undue fatigue on the part of the operator using a lever with a movement similar to **Fig 7.5.** These data represent good design parameters for lever size and load for all types of handpumps operated by lever action.

Typical performance achieved with wing type (semi-rotary) handpumps is summarised in **Table I.** Other data are given in **Tables II** and **III.**

TABLE I – TYPICAL SIZES OF SEMI-ROTARY HAND PUMPS (Max Suction Lift Approx 20 ft (6.5 m)

Size (Suction & Delivery)	½ in 12.5 mm	¾ in 20 mm	1 in 25 mm	1¼ in 32 mm	1½ in 46 mm	2 in 50 mm	2½ in 65 mm	3 in 75 mm
Double acting gal/hr	160	265	375	540	800–900	1500–1700	2100	4000
lit/min	12	21	28	40	66–67.5	110–130	160	300
Quadruple-acting gal/hr	–	460	600	840	1200–1500	1850–2000	–	–
lit/min	–	39	45	63	90–110	140–150	–	–

TABLE II – TYPICAL SIZES OF DIAPHRAGM HAND PUMPS

Size		gal/hr	lit/min	Speed
2 in	(50 mm)	800–1000	60–79	40 double strokes/min
3 in	(75 mm)	2000–3000	150–225	40 ” ”
Twin 3 in	(75 mm)	5000–6000	375–450	40 ” ”
4 in	(100 mm)	4000–6000	300–450	30 ” ”
Twin 4 in	(100 mm)	8000–12000	600–900	30 ” ”

TABLE III

PERFORMANCE OF ECCENTRIC SCREW HAND PUMP (Monolift)

Delivery : up to 600 gal/hr (45 lit/min)
Heads : up to 200 ft (60 m)
Typical performance (at 30 ft (10 m) head) : 250 gal/hr (20 lit/min) at 50 handle turns (150 rotor turns) per minute.

8. Miscellaneous Pump Types

NUMEROUS other pump types exist which are original designs in the sense that they work on known principles not commonly used in other pumps; or alternatively are basically standard forms of pumps differing appreciably in detail design, with claimed advantages. The following is only a partial list, as 'new' pump designs will continue to emerge, although it appears highly unlikely that any new *pumping principle* itself will be discovered. The main pump developments, in fact, are in materials, improvements in hydraulic performance and pumping efficiencies (particularly in the case of centrifugal pumps); and detail changes in standard designs of pumps to adapt them for handling a wider range of fluids.

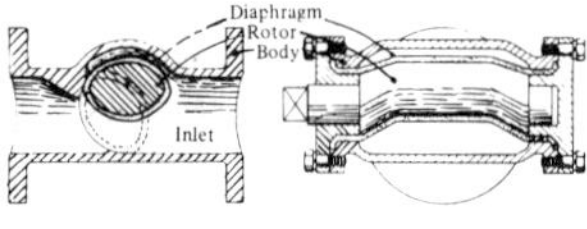

FLEXIBLE DIAPHRAGM

This type employs a tubular diaphragm associated with an eccentric rotor producing a 'rolling squeeze' action; or surrounded by a liquid chamber in which the fluid is subject to pulses of hydraulic pressure to provide the squeeze action.

Although referred to as a 'diaphragm' pump it is more truly a peristaltic pump with the advantage that large tube sizes can be used. It is a glandless pump that can handle corrosive fluids as well as solids in suspension.

Another form of this pump is where the flexible tube is pulsed directly from a compressed air supply.

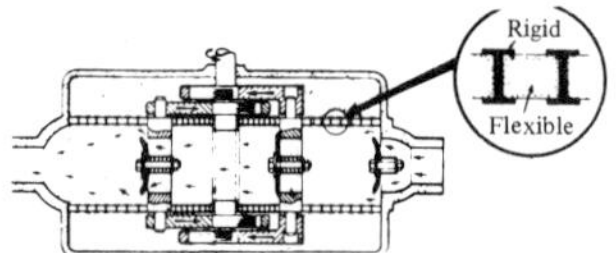

STACKED DIAPHRAGM

The stacked diaphragm pump comprises stacks of alternating rigid and flexible chamber rings with interposed valve assemblies dividing the length on to separate volumes (three is a typical number). An eccentric (mechanical) drive is applied to the valve sections which then alternately move towards and away from each other, this movement being accommodated by the flexibility of the chamber walls. Through-pumping action is thus promoted in conjunction with a valve on the inlet side to eliminate back-flow.

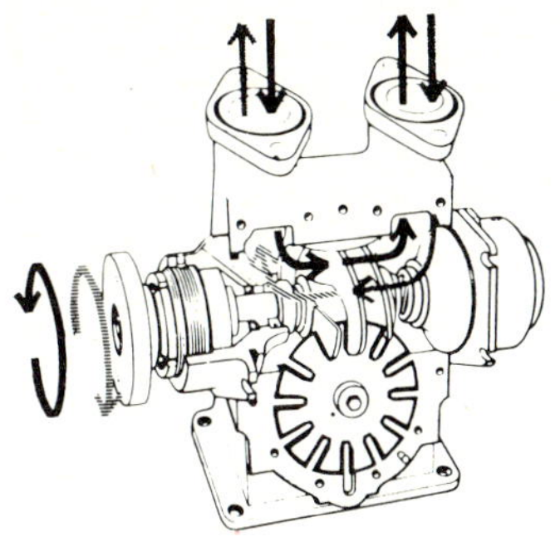

GOODYEAR PUMP

This pump employs a short, large diameter helix form impeller engaging with a slotted elastomeric disc in such a manner as to provide throughput with either direction of rotation. No valves are needed and it can handle a wide range of both clean and contaminated fluids.

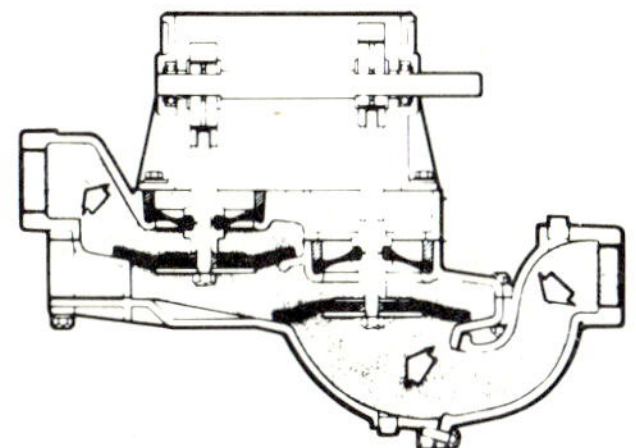

DOUBLE DISC PUMP

In the Double Disc pump a rotating shaft reciprocates two rubber discs in opposition to each other, thus providing an expanding/contracting volume between them to produce positive displacement pumping action without the need for valves (the discs themselves act as inlet and discharge valves). The design is also double-acting and, with generous through passages, can handle a wide variety of products. It will also dry prime and, when primed, has a high suction lift capability. Practical sizes are limited to capacities of above 200 gal/min (900 lit/min).

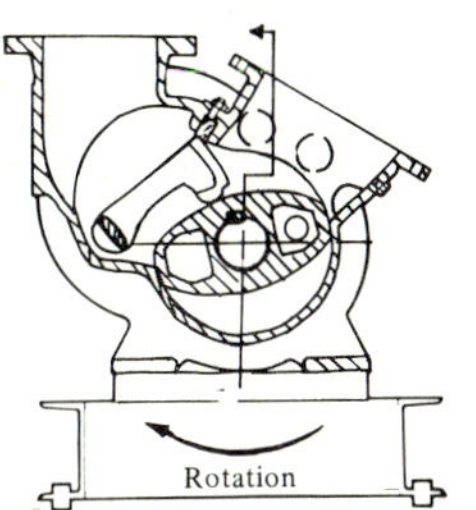

ELLIPSE AND SCRAPER

Pump based on an elliptic rotor rotating in a cylindrical casing with a hinged sealing arm or scraper bearing on the surface of the rotor. Capable of handling highly viscous products in a gentle manner, particularly those of a crystalline nature where minimum crystal breakdown is required. Also known as a *massecuite pump*. Pumps of this type are hopper fed and delivery pressures are normally less than 100 lb/in^2 (7 bar).

INCLINED ROTOR

A rotary pump where the impeller is mounted obliquely on a shaft rotating in a circular casing.

TORQUE FLOW

A rotary pump with an impeller in the form of a notched disc, pumping pressure being developed by torque reaction.

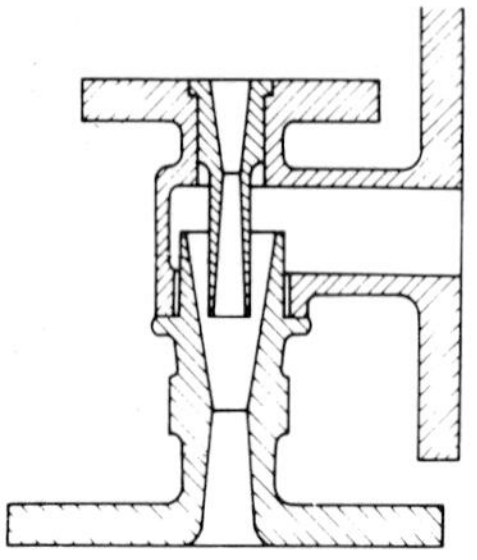

EJECTORS

Ejectors (and injectors) are simple pumps with no moving parts, operating via a pressurised fluid (normally steam) expanding in a convergent/divergent nozzle, producing a low pressure high velocity jet. This evacuates air in the suction pipe, drawing fluid into the ejector which then flows out with the jet stream. Water is used instead of steam for ejectors used to exhaust air (*ie* as a vacuum pump).

Ejectors may be used for lifting, forcing, or a combination of both. Other examples are *borehole ejectors* for water raising, and compressed air ejectors for exhausting gases (but not liquids).

DRUM-ROTOR

Variant of the lobe rotor pump with slotted drum and vaned drum intermeshing. Capacity directly proportional to speed (less slippage). Low head (usually). Not self-priming.

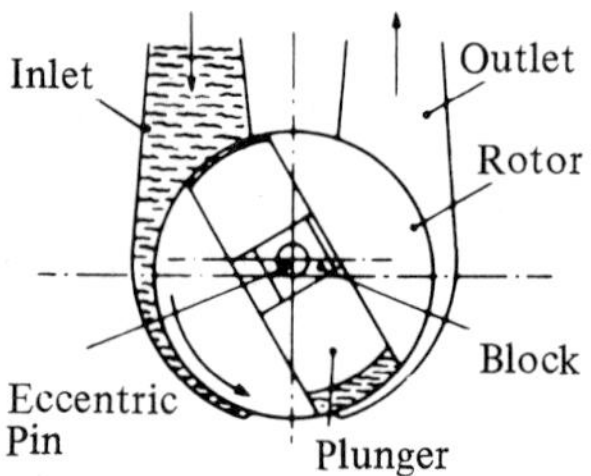

SHUTTLE-BLOCK

Simple form of rotary piston pump capable of handling clean fluids of low to moderate viscosity. Size range restricted. Capable of developing pressures up to about 75 lb/in^2 (5 bar).

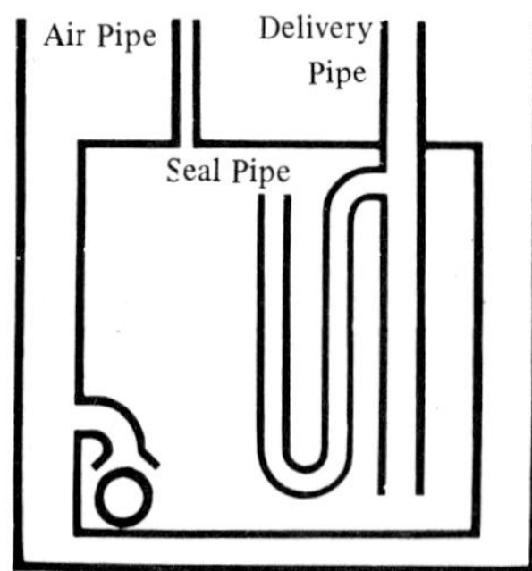

'PNEU'

Pump with no moving parts except for a single outlet valve, capable of handling a wide variety of liquids ranging from water to sludges, slurries and viscous fluids. Works off compressed air supply.

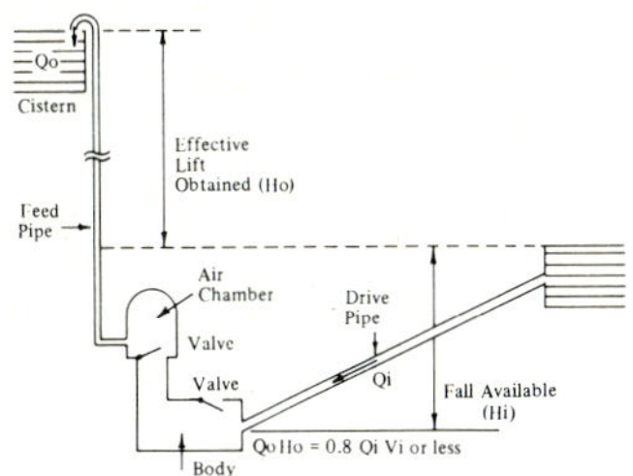

'HYDRAM'

Hydraulic ram pump with no moving parts other than two flap valves utilising energy recovery from a fall of water to raise a proportion of that water to a greater height. Capable of working at up to 80% energy recovery – *eg* 100 gallons (457 lit) of water falling through 10 ft (3 m) is capable of raising 10 gal (46 lit) of water 80 ft (25 m).

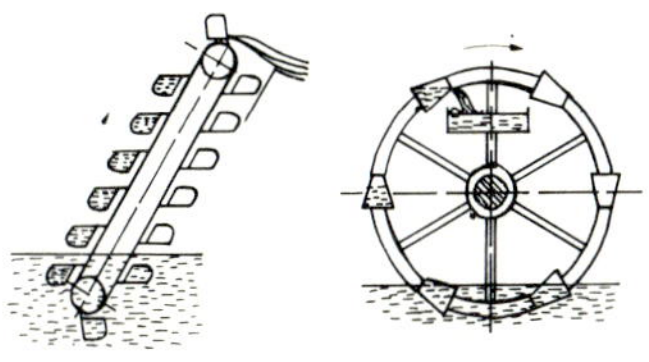

BUCKET

Elementary form of mechanised water pump for low lifts; further developed as a chain pump or bucket elevator for higher lifts.

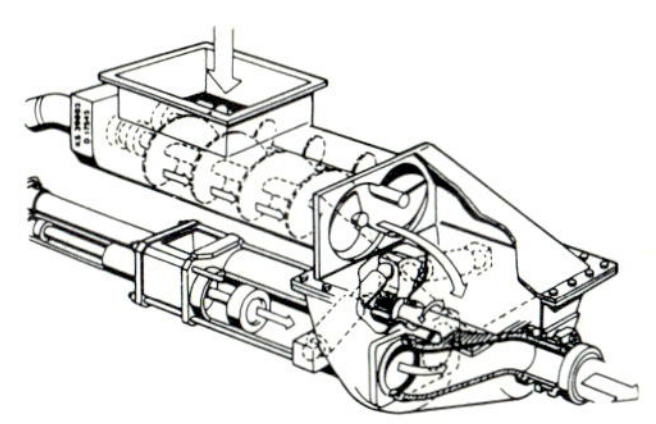

RAM PUMP

Special type of plunger pump for handling clays, muds, thick slurries, *etc* which cannot be inducted into a cylinder in the normal way. The product is fed via a hopper into a cutting cylinder which can be advanced into the chamber to be filled. The chamber is then discharged by entry of a ram. Both cutting cylinder and ram motions are commonly driven by hydraulic pressure.

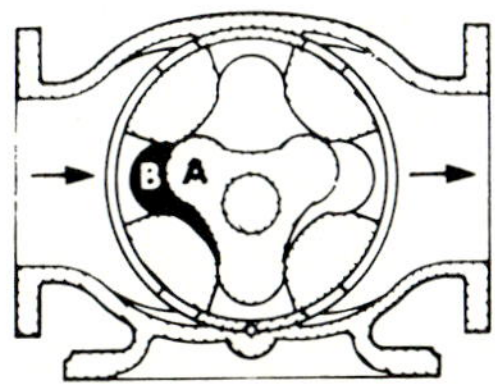

INTERNAL LOBE ROTOR

Like the gear pump, the lobe rotor pump also has two possible configurations – 'external' and 'internal'. In the internal lobe rotor pump one lobed rotor is rotated synchronously within a lobed external ring normally having one additional lobe.

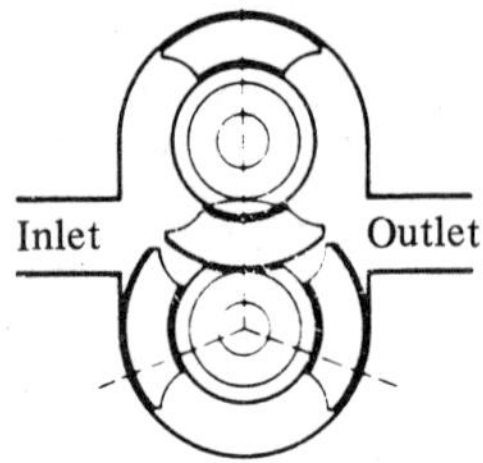

'WAUKESHA'

This, basically, is another form of lobe rotor pump. In this case instead of employing rotor forms pumping action is produced by arc-shaped (curved) pistons traversing annular-shaped 'cylinders'. Although employing pistons as such it is a true rotary pump.

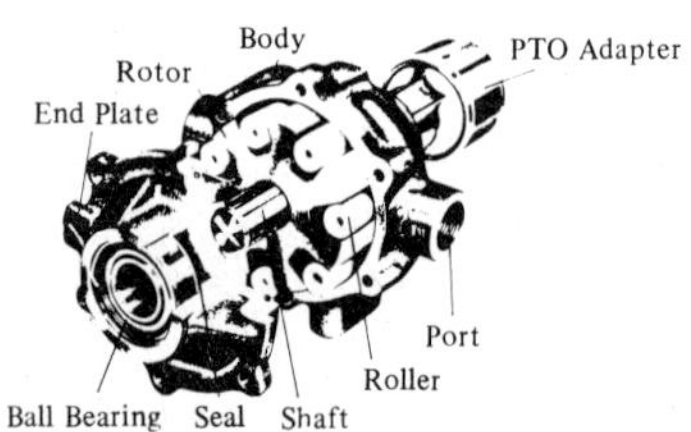

ROLLING VANE

The rolling vane pump employs a grooved, eccentrically mounted rotor carrying elastomeric rollers instead of vanes. It is capable of handling fluids containing small solids (like a flexible vane pump) whilst retaining an ability to deliver high pressures like a rigid vane pump.

9. Products Handled

ROTODYNAMIC PUMPS, in general, are mainly used for the handling of clean, cold water and aqueous solutions. This does not necessarily preclude their use with more viscous clean fluids, or the handling of liquid-solid mixtures with suitable designs of impellers. The performance of such pumps is, however, almost always determined when handling clean, cold water and the water performance may be considerably modified when handling other classes of fluids. Equally, there are limits set by the properties of non-aqueous liquids which may make the choice of a rotodynamic pump unsuitable, or uneconomic.

Many other pumps are suited to handling a wider range of fluid viscosities, although again there may be distinct limitations where solids in suspension are involved. Performance will still be affected by fluid viscosity, but not necessarily in the same manner or to the same degree. Thus positive displacement pumps, in general, are capable of handling a very wide range of fluid viscosities, provided the speed is adjusted to maintain satisfactory efficiency levels. Also performance may be quoted for one or more specific fluids or fluid viscosities rather than water performance (*eg* gear pump performance is commonly specified for oils as the 'standard' fluid).

Fluid characteristics, therefore, determine:

(i) The suitability or otherwise of a particular pump type for handling that that fluid.

(ii) Preference for the selection of a particular pump type where the fluid characteristics are particularly 'difficult'.

(iii) The actual performance of the pump when handling that fluid.

(iv) Modifications to the detail design which may be desirable or necessary for most effective working. Note here that 'effective working' includes both efficiency of working and consistency of operation.

These factors largely influence the selection of a suitable pump type and size. Various other factors may then have to be taken into account – such as the chemical nature of the fluid affecting material selection – but these are secondary to type selection unless the fluid is highly corrosive. In many cases several alternative types may be suitable, when final selection can be based on the other practical factors involved and overall economics.

Aqueous Fluids

Clean aqueous fluids normally have a viscosity and density similar to that of water and thus the 'water performance' of the pump generally applies for both type selection and performance. Selection requirements may be modified, however, if the fluid is chemically active.

Normally this will mainly affect material selection or call for the use of a *chemical pump.* It can also dictate the preferred type of pump seal(s), particularly if leakage is to be contained.

A point which should not be overlooked is that certain aqueous solutions, notably brines, may have an appreciably higher density than water and the power input required by the pump will be increased proportionately; and also the frictional losses, affecting total head. With saturated brines the specific gravity may reach a figure as high as 1.3.

Viscous Fluids

Viscous fluids require more power to pump, or be pumped at lower speeds to maintain acceptable power input levels. The problem is aggravated if there is a marked change in viscosity during handling, *eg* the pump is started up with a cold, very viscous product which becomes much less viscous at normal system temperatures.

A method of overcoming this is to heat the product container before starting and so render it fluid at normal pumping temperature. Product and/or pump heating can also prove the most economic method of handling a variety of viscous fluids, and, if necessary, heating can also extend to pipes and valves.

Pumps designed for heating normally incorporate a double casing forming a hollow jacket through which a suitable heating fluid can be circulated, *eg* hot water or steam. Standard designs of pumps can be heated by covering with an electric heater in blanket form. Blanket or tape types of electrical heaters can also be used for jacketing pipes and valves, although in larger pipe sizes double walled construction can be used with heating fluid circulated between them.

Unless there are very definite energy or economic savings, however, or the process itself benefits from heating the products, most viscous fluids are handled by standard pumps at the normal fluid viscosity.

Viscous Fluids – Centrifugal Pumps

Centrifugal pumps have a limited application for handling fluids of this type since an increase in fluid viscosity will increase the power input required and at the same time derate the H-Q curve. The overall result is a marked decrease in efficiency with increasing fluid viscosity, as shown in the generalised diagram of **Fig 9.1**.

More specifically, but still in general terms, with fluids of relatively low viscosity (up to about 20 centistokes) centrifugal pump performance is largely

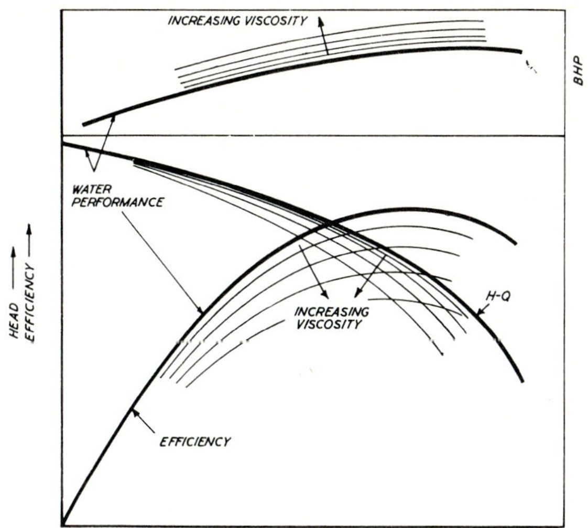

Fig 9.1: Typical characteristic curves for a centrifugal pump handling water and more viscous fluids. Note the rapid loss of efficiency which generally starts to show at fluid viscosities in excess of 100 centistokes.

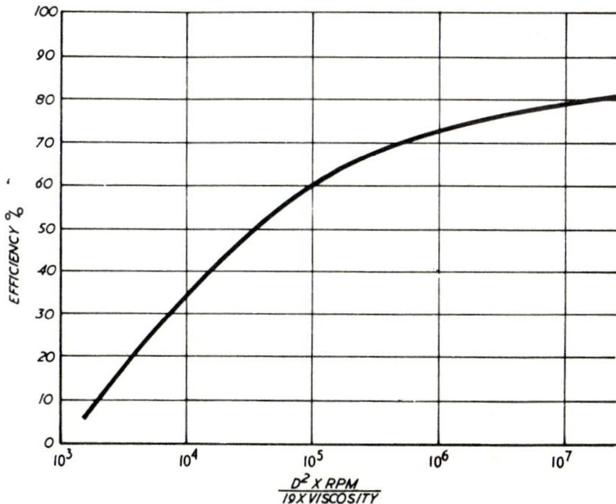

Fig 9.2: Design parameter for a centrifugal pump 'sized' to handle a viscous fluid. A high relative efficiency can be obtained only with large impeller diameters and relatively high operating speeds.

maintained on the same level as water performance. In the viscosity range 20–100 centistokes there is a small but progressive reduction in head (*ie* a lowering of the H-Q curve) but an appreciable reduction in efficiency because of the higher power input required. With fluid viscosities above 100 centistokes there is a marked loss of head and a considerable reduction in efficiency, although the latter will depend very largely on the impeller diameter and speed or pump Reynold's number (R_e).

Size effect is shown in **Fig 9.2** where relative efficiency is plotted against R_e (calculated in feet second units). Such a curve will be specific to particular design of pump, but the general form of the curve will usually be similar for all cies when handling viscous fluids the pump needs to be as large as possible and

Brooks centrifugal pump for handling corrosive viscous and non-viscous fluids.
Delivery – unfiltered waste materials to storage tank above
Materials delivered: 100% solution caustic soda
80% solution caustic soda
Benzine
Alpha Picolene
Vinyl Pyridene

be run at the highest possible speed. Thus where large capacities are involved a large centrifugal pump can show quite high efficiencies when handling oils and similar liquids of moderate to moderately high viscosity (say up to 3 000 or even 5 000 centistokes). Small centrifugal pumps on the other hand are basically unsuited to handling viscous liquids because of the low efficiencies resulting and are virtually not worth considering for such duties when the fluid viscosity exceeds 100 centistokes. Other types of pump can provide low capacity demands at much higher efficiencies.

The performance of a typical 2 in (50 mm) centrifugal pump handling water and oils of various viscosities is shown in **Fig 9.3**. In addition to a steepening and lowering of the H-Q curve the point of optimum efficiency occurs at decreasing values of Q with increasing viscosity, this being a general characteristic. In some cases, however, the form of the H-Q curve may also be modified, so that optimum efficiency with more viscous fluids may occur at even lower values of Q than indicated by the typical pattern. This is largely dependent on the clearance between the impeller and casing, a very small clearance tending to boost the head realised by smaller values of Q.

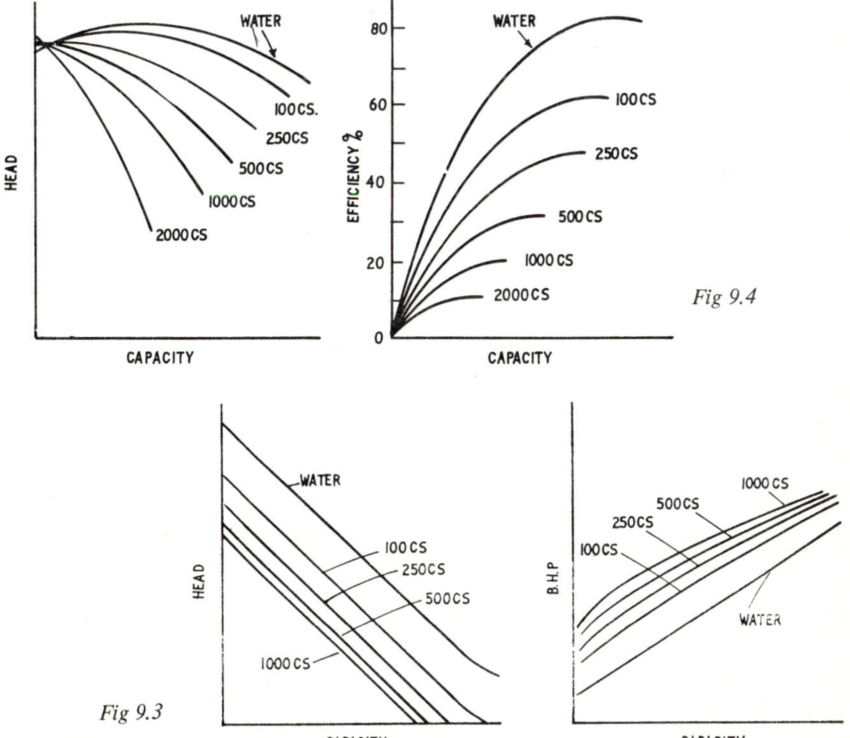

Fig 9.4

Fig 9.3

Regenerative type pumps are even more affected by increasing fluid viscosity and come into the same category as small centrifugal pumps – *ie* they are generally unsuited to handling liquids with a viscosity greater than 100 centistokes, except at low efficiencies – see **Fig 9.4**.

The performance obtained from a centrifugal pump handling a viscous liquid can be estimated by applying a correction factor to the water performance. Basically, for a constant speed the value of the specific speed does not change for a given value of power input, whence a relationship between Q_w, Q_l, H_w and H_l can be derived (the suffices referring to w = water and l = viscous fluid).

$$\frac{Q_w}{Q_l} = \left(\frac{H_w}{H_l}\right)^{1.5}$$

Suitable correction factors can only be determined experimentally by direct measurement of loss of characteristics and then analysing the data obtained. If done with a specific pump, specific curves for different liquid viscosities can be presented, as in **Fig 9.5**. With sufficient data from numerous pumps tests a general pattern of behaviour can be established to yield correction factors applicable to any centrifugal pump to correct its water performance.

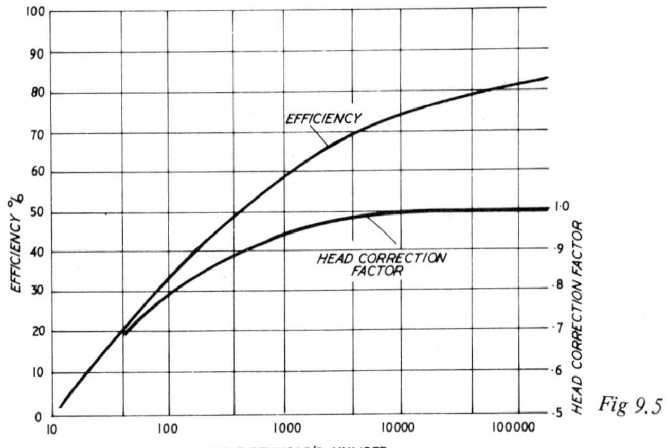

Fig 9.5

A certain difficulty arises in the parameters to be adopted for presentation. Any form of correction factor chart tends to become extremely complex if presented directly in terms of head, capacity and efficiency, although this has been done and such charts are available (US Hydraulic Institute). A rather more straightforward analysis of experimental data is to derive correction factors related to a single parameter, for which the Reynold's Number is a logical choice as including flow, liquid viscosity and pump size (impeller diameter). For convenience, however, it is better to calculate this particular Reynold's Number in terms of discharge (Q), *viz*

$$R_e = \frac{Q}{\nu D^2}$$

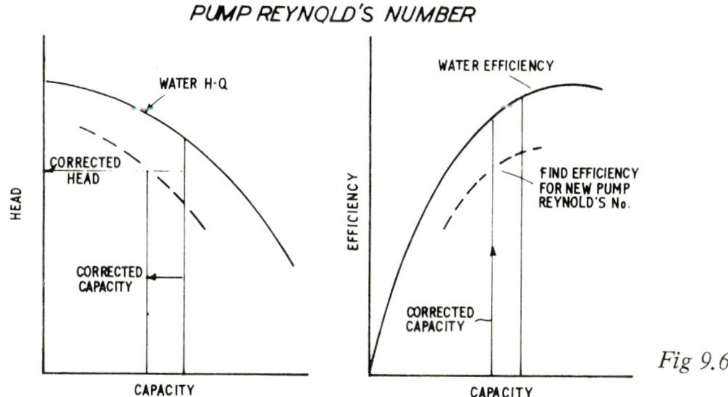

Fig 9.6

From available data it is then possible to plot correction factors against R_e typical curves of this type being shown in **Fig 9.6** (Stepanoff). These curves give correction factors only for head and efficiency, but enable the full characteristics to be corrected following a typical procedure, *viz:*

(i) Calculate R_e for a nominal value of Q and the liquid viscosity and pump size.

(ii) Enter graph with calculated value of R_e to determine correction factors for head and efficiency.

(iii) Correct head figure.

(iv) Use head correction factor to calculate corrected capacity figure from $Q_w/Q_e = (H_w/H_e)^{1.5}$ and correct capacity. This establishes the derated H-Q point.

(v) Correct efficiency and establish corrected efficiency point.

This procedure can be repeated with different starting values of Q to replot the H-Q and efficiency curves, showing the complete performance of the pump handling liquid of viscosity ν.

The power input required can be calculated directly from the corrected values of Q_1 and H_1 and η_1, either at a specific point or as a series of calculations to establish the power input curve.

$$\text{Power input HP} = \frac{QHw}{K\eta}$$

where

w = specific weight of the liquid.

η = pump efficiency at H-Q point considered.

K = a constant depending on the units employed for Q_1, H and w in gal/min.

= 51.5 for Q in (Imp) gal/min, H in feet and w in lb/cu ft.

= 62.0 for Q in (US) gal/min, H in bar and w in lb/cu ft.

= 71.5 for Q in lit/min, H in metres, and liquid weight expressed as SG.

An alternative method of working is to estimate a suitable size of pump (based on water performance) and find the correction factor for head based on the required delivery Q_r. From this determine the correction factor for capacity = (head correction factor)$^{1.5}$. Divide Q_r by the capacity correction factor; and divide the required head by the head correction factor. These two new figures for capacity and head will give the required working point in terms of water performance on which a final choice of a suitable pump size can be made.

Corrections to water performance applied in this way are not necessarily exact since the factors represent means of empirical data and individual designs vary in performance by as much as 10% (although the spread is usually lower). Also correction factors can only be applied to centrifugal pumps of conventional form; and for uniform homogeneous liquids. The method may, however, be extended to liquids containing solids where the mixture is homogeneous and thus the flow characteristics closely approximate to that of a viscous fluid.

Viscous Fluids – Reciprocating Pumps

Reciprocating pumps, in general, are capable of handling liquids of up to about 100 centistokes viscosity without any reduction in speed and capacity, although the efficiency will be slightly lower than with water performance because of the increase in power input required. At higher viscosities, efficiency is likely to fall still further because of increased resistance to flow through ports and valves. It is thus usual to operate reciprocating pumps at reduced speed when handling higher viscosity fluids. Typical recommendations are shown in **Fig 9.7.**

These values are illustrative only, as much depends on the size and detail design of the pump (particularly valving) but above about 2 000 centistokes viscosity permissible speed of pumping becomes increasingly slower. It is fairly common practice, however, where reciprocating pumps are used to handle higher viscosity fluids that an 'oversize' pump is selected to run at a reduced speed. Thus for handling a fluid with a viscosity of around 1 000 centistokes a pump

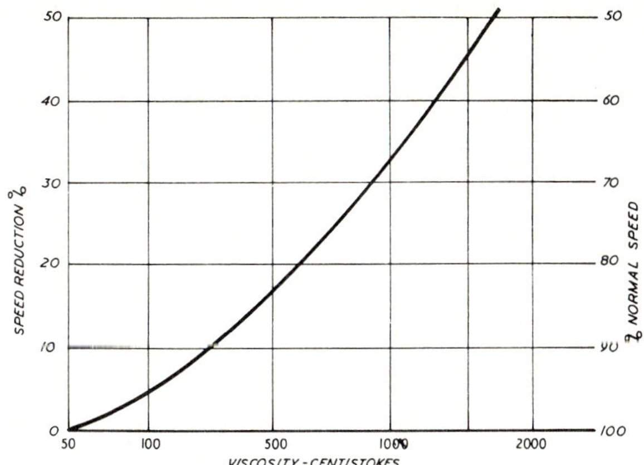

Fig 9.7: Typical speed reduction requirements for a conventional reciprocating pump handling viscous fluids. Once the fluid viscosity exceeds about 1 000 centistokes quite drastic speed reduction is usually necessary.

with three times the required capacity would be chosen and run at approximately one third the design speed. At high viscosities it may also be desirable, or even necessary, to design the pump to operate without a suction valve and modify the discharge valves to ensure free flow. By this means it is possible to extend the application of the reciprocating pump to handling the most viscous fluids and semi-solids, the ultimate in this respect being the hydraulic ram pump with intake via a hopper feed rather than suction and a very low rate of stroking – see also **Fig 9.8.**

Reciprocating pumps are also particularly suited for high pressure duties. Rotary piston or rotary plunger types are usually preferred for small capacity high pressure services involving oils or hydraulic fluids.

Viscous Fluids – Rotary Pumps

Rotary pumps are generally capable of handling a wide range of liquid viscosities and standard models of certain types may be capable of handling viscosities of up to 50 000 centistokes without modification, simply by adjustment of speed.

Gear pumps, in particular, are generally regarded as most suitable for handling oils and similar fluids possessing lubricating properties. Since these pumps operate with small but positive clearances a certain minimum fluid viscosity is also required in order to maintain adequate sealing and thus minimise internal losses.

The simple external gear pump is particularly suitable for handling lubricating fluids with a minimum viscosity down to 10 centistokes (with close clearances). For any given size, capacity is directly proportional to speed, decreasing slightly

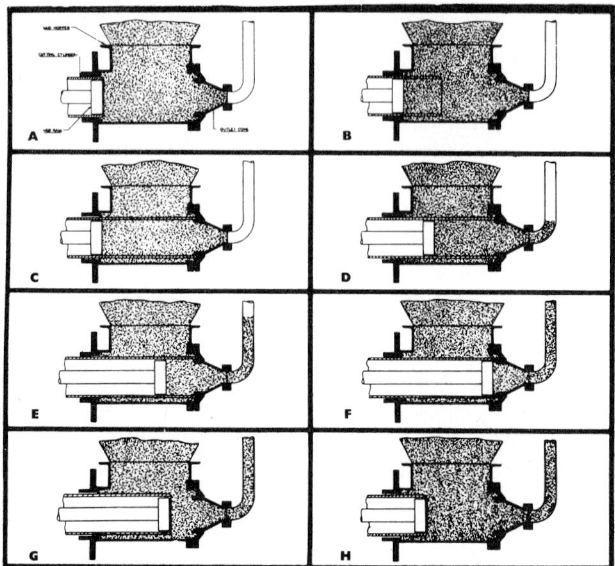

Fig 9.8: Sequence of operation of Meader ram pump for handling clays, muds, thick slurries, etc.

A. The chamber is loaded with material
B. and C. The cutting cylinder forces its way into material
D. E. and F. The ram forces the material from the cylinder into the discharge pipe
G. and H. Cylinder and ram together are withdrawn and further material falls to bottom of chamber.

at any given speed with increasing head, since slip increases with head but is substantially independent of speed. The effect of increasing fluid viscosity is to decrease slip, but increase the power input required to maintain speed (and thus capacity). Efficiency thus tends to decrease with increasing fluid viscosity. In practice, therefore, it is general to specify or recommend a reduction in operating speed (and capacity) with increasing fluid viscosity for a given type and size of pump **(Fig 9.9)**. The detail design of the pump may also be modified for specific viscosity ranges (*eg* increased clearances for higher viscosities to reduce power input rise) and also for different delivery pressures (high pressure gear pumps requiring greater attention to minimising leakage across the ends of the gears). Inter-tooth venting may also be necessary when high fluid viscosities and/or high fluid pressures have to be accommodated. An alternative method of rotor tooth relief when handling high viscosity fluids is to reduce the number of gear teeth employed. Single helical gearing is normally preferred both to produce low shear levels in the product and give good self-cleaning of the gears themselves.

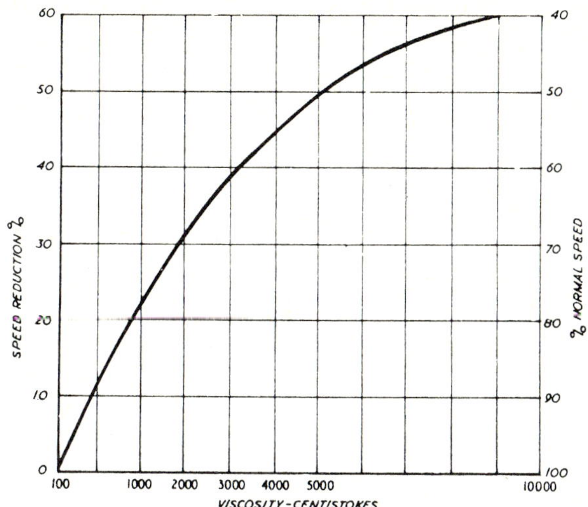

Fig 9.9: Typical speed reduction requirements for rotary pumps. In this case initial speed reduction is greatest and quite high fluid viscosities can often be accommodated without drastic speed reduction. Many standard rotary pumps are capable of handling fluid viscosities well in excess of 10 000 centistokes, although clearances may have to be increased (particularly with lobe rotor pumps).

Fully heated (*ie* water jacketed) external gear pumps are particularly suitable for handling plastic melts of medium to high viscosity, as well as other more viscous products. Intake pressure feeding can substantially increase the viscosity of the product which can be handled.

Internal gear pumps are also suitable for handling viscous fluids but pumps designed specifically for this type of product normally have lobe-shaped rotor forms with a minimum number of teeth as this configuration can materially improve volumetric efficiency.

Viscous Fluids – Lobe Rotor Pumps

The lobe rotor pump operates with small but positive clearance and can thus handle fluids which do not possess lubricating properties. It is generally suited for handling all types of clean fluids up to a viscosity of about 250 centistokes without modification, although it may be necessary to reduce the operating speed with increasing viscosity to minimise cavitation (reduce noise and vibration) and maintain good efficiencies. A limiting speed is usually set where the inflow breaks down or channels and the higher viscosity liquid no longer fills the pockets between the lobe fully, causing the pump to chatter.

Increasing the clearances has a beneficial effect for handling higher viscosity liquids. Any increase in slip resulting is actually beneficial in that it provides a certain amount of damping and reduces channelling tendencies. As a general rule, however, the greater the liquid viscosity, the less the slip, even with quite generous clearances, setting fairly clearly-defined empirical limits to operating speed.

Whilst gear pumps are the primary choice for handling oils where relatively low capacities are required, lobe rotor pumps are more suitable for production in larger sizes and capacities and yield higher specific outputs for a given size of pump. They are thus a preferred type in the low-medium to medium capacity range, and can handle a wider range of viscosities than gear pumps. The rotor form can also be 'tailored' to suit the characteristics of the fluid being handled, *eg* to provide a gentle pumping action if the product being handled is shear sensitive.

Viscous Fluids – Vane Pumps

The simple vane pump is generally regarded as a light-duty type for use with oil fluids up to moderate pressures. The sliding action of the vanes is impaired and the slip increased when higher viscosity fluids are involved and thus the simple vane pump is usually restricted to use with fluids of lighter viscosity.

The same principle is employed in heavy-duty rotary-vane pumps which have proved capable of handling the most aggressive as well as the most viscous fluids, semi-solids and even solids. In this case a substantial one-piece single vane only is employed, sliding in a slot in an eccentric rotor, the internal components of the pump being protected, where necessary, by hard-facing to handle abrasive substances. Pumps of this type have the ability to handle almost any substance that can pass into the casing, with the speed of the pump adjusted accordingly. In exceptional circumstances pumping speed can be as low as one revolution per hour whilst still maintaining high volumetric efficiency. Speed is the controlling factor as regards overall efficiency (*ie* the speed is selected to arrive at an optimum power input for the product being handled).

Viscous Fluids – Screw Pumps

Screw pumps have a versatile application in the handling of fluids over a wide range of viscosities. Basically there are two main types:

(i) single-screw pumps which have a spirally formed rotor which rotates eccentrically in an elastomeric stator;

(ii) parallel screw pumps where two or more rigid screws of opposite pitch rotate with close clearances.

Both types operate as positive displacement machines, producing a virtually pulse-free axial flow along the length of the body. The capacity of single screw pumps is generally limited to about 250 gal/min (55 lit/min), although such pumps can, if necessary, be extended in length or mounted in tandem, and normally operate at relatively low speeds. They have a particularly wide range of

Eccentric screw pump used for transporting fruit jams to the filler heads of jar filling machines. Pump speed 28–30 rev/min to fill approximately 2 000 jars per hour.

application for handling chemicals, foodstuffs and similar products and are well suited to the handling of shear-sensitive fluids and delicate semi-solids. The basic layout also lends itself to adaption for the most effective handling of specific fluids, *eg* by selecting a suitable material and hardness for the elastomeric stator.

Rigid screw pumps employ two, or sometimes three, axial screw rotors meshing together with fine but positive clearance and timed by gears. To balance end thrust each rotor commonly comprises two separate opposite-hand screws so that the complete unit has centre feed and discharge from each end. Machines of this type are usually larger than single screw pumps and may run at the same level of rotational speeds as centrifugal pumps, although axial displacement of the fluid takes place at comparatively low speeds. Standard pumps of this type may, however, have capacities in excess of 1 000 tons per hour for handling petroleum spirit and similar low viscosity fluids and fuel oils with viscosities up to about 500 centistokes. More viscous fluids are generally handled at lower rotational speeds and thus reduced capacity for a given pump size.

Since such pumps operate with positive clearance at all times they are suitable for handling low-viscosity fluids which lack lubricating properties as well as definite lubricants and more viscous fluids. Where the fluid is a lubricant, bearings and timing gears can be mounted internally rather than externally and lubricated by the fluid being handled rather than independently lubricated. Clearances may be selected according to the fluid being handled, and the duty involved. Thus for high-pressure working with lubricants the screws may actually contact the body bores to eliminate internal leakage (normally on smaller sizes of screw pumps only), whilst quite generous positive clearances may be used for handling more viscous fluids such as molasses. An outstanding characteristic of rigid screw pumps is their suitability for high-pressure services, particularly where large capacities are required.

Pumps for Oil Fluids

The selection of the most suitable pump for a service involving a low to medium viscosity oil affords an excellent example of the influence of specific requirements on final choice. Quite obviously, from the previous descriptions a variety of pump types are suitable for the basic duty, of which one or more may be rejected as unsuitable by size (*eg* a small centrifugal pump would be the least preferred for small capacities). Thus:

(i) Where the capacity requirements are small a gear pump will probably provide the simplest and most economic solution. A suitable size can then be selected on the basis of speed (and thus capacity) recommended with the oil viscosity concerned.

(ii) Where a rather higher capacity is required a lobe rotor pump will probably be the best choice. This type will also extend the capacity available outside the normal range of gear pumps.

(iii) If the capacity required is large –

  (a) A centrifugal pump will probably be the optimum choice if the oil viscosity is not greater than 100 centistokes.

  (b) If the fluid viscosity is greater than 100 centistokes, choice may lie between a centrifugal, reciprocating or rigid screw pump. The respective merit of each should be assessed in detail, with a final choice based on overall economics (unless other special requirements overrule).

(iv) If a high pressure output is required, compare capacity requirement against type, *eg*

  (a) Small-moderate capacity – rotary plunger or rotary piston.

  (b) Moderate or high capacity – reciprocating or rigid screw.

(v) If the fluid is contaminated, consider

  (a) Flexible vane pump for light duties.

  (b) Lobe rotor pump for medium duties.

  (c) Screw pump for large capacities (or possibly centrifugal for low viscosity oils).

Solids in Suspension

The presence of abrasive solids in suspension in the liquid being handled limits selection to pump types which can handle such solids without jamming or clogging, or causing excessive wear on the pump components. Certain pumps are inherently capable of passing such solids, or are capable of so doing with modification to detail design. Other types are inherently incapable of handling other than clean fluids and are thus excluded from selection for such services. These characteristics are dealt with elsewhere under individual pump headings, but see also **Table I.**

The presence of soft solids calls for a pump having clearances suitable for passing such solids, and the elimination of voids in which such solids might collect. If the solids are of a delicate nature, selection may be further limited to a pump type which can pass such solids without damaging them.

All solids which are suspended in the fluid rather than being intimately mixed and dispersed form non-homogeneous mixtures. Flow is thus heterogeneous, with a tendency for particles to slide at the surface of a more or less stationary bed formed by settlement at very low velocities; or more in the manner of a sliding bed at slightly higher velocities; or be retained fully suspended in the fluid stream at velocities higher than the settling velocity.

The effect on pump performance will be generally to derate the water performance, although the full water performance may nominally be achieved at low to moderate concentrations. Thus the water performance is commonly employed to select a suitable size of pump, with a preference for 'oversizing'. This has relatively little significance except that in the case of a centrifugal pump efficiency may suffer if greatly 'oversized' and the required capacity may not always be realisable if an exact size is selected on water performance.

Estimation of frictional losses in the pipework is also difficult, and thus determination of total head. Typical values can only be determined by experiment or test, although a semi-empirical formula (Worster and Durant) which gives results consistent with practice for coal, gravel, sand and similar solids in suspension is:

$$\frac{\Delta P_S}{\Delta P_W} = K.c. \frac{D(Sg - 1) \times V_S}{\sqrt{d\,(Sg - 1) \times V^2}}$$

where:

ΔP_S = pressure drop when transporting solids

ΔP_W = pressure drop when transporting water

c = concentration of solids by volume

S_g = specific gravity of solids

V_S = settling velocity of solids

V = flow velocity

K is an empirical factor

This holds good provided that the flow velocity is greater than the settling velocity of the heterogeneous mixture.

TABLE I – PUMP SUITABILITY FOR HANDLING DIFFERENT FLUIDS

PUMP TYPE	CLEAN FLUIDS				CONTAMINATED FLUIDS				Pulps	Pastes and Greases	Sensitive Fluids
	Water and Aqueous Solutions	Oils	Viscous Oils, etc	Very Viscous	Soft Solids	Stringy Solids	Abrasive Solids	Delicate Solids			
CENTRIFUGAL											
Standard	E	S	L	X	L	X	X	X	X	X	X
Lined	E	S	L	X	S	–	S	–	–	–	X
Chokeless	L	L	L	X	S	S	S	L	S	L	X
Bladeless	–	–	–	X	S	S	L	E	L	–	X
Self Priming	E	S	–	X	L	–	–	–	–	–	X
Multi-stage	E	S	–	X	–	–	–	–	–	–	X
Submersible	E	S	–	–	E*	S*	S*	–	S*	–	–
REGENERATIVE	E	S	X	X	X	X	X	X	X	X	X
MIXED FLOW	E	S	–	–	L	–	–	–	–	–	–
AXIAL FLOW	E	S	–	–	L	–	–	–	–	–	–
RECIPROCATING											
Piston	E	E	S	S	L	L	L	X	L	L	X
Plunger	E	E	S	L	–	–	X	X	L	L	L
Radial Piston	L	E	S	L	–	–	X	X	–	–	–
ROTARY											
Gear	L	E	E	S	L	X	X	X	X	S	X
Lobe Rotor	L	E	E	S	S	L	L	S	S	S	S
Vane	S	E	S	L	–	X	X	X	L	L	L
Heavy-duty Vane	–	–	–	–	E	L	E	L	S	S	S
Single Screw	S	S	S	E	E	E	E	E	E	E	E
Multiple Screw	S	E	E	S	–	–	–	–	–	–	–
ROTARY											
Flexible Vane	E	E	L	X	S	L	L	X	L	L	L
DIAPHRAGM	E	E	S	L	E	S	E	S	S	S	S
PERISTALTIC	S	S	S	S	E	E	E	E	E	E	E
EJECTORS	S	L	X	X	X	X	X	X	X	X	X

E – Excellent S – Suitable L – Limited Suitability X – Unsuitable *with smaller impeller

Note: No entry (–) does not necessarily indicate lack of suitability, but rather that the type of pump concerned would not normally be a first choice for that particular application unless other circumstances influence choice.

Slurry pump type LPN-250/600. (Serlachius Pump Factory)

Something like 95% of abrasive slurries pumped are handled by centrifugal pumps (slurry pumps), with diaphragm, single screw and vane pumps sharing the remainder (with individual exceptions).

Centrifugal slurry pumps are of special design, the primary requirements being suitable resistance to abrasion and effective shaft sealing. They are thus generally designed around suitable wear-resistant materials and special types of seals.

In general, lined pumps (*eg* rubber lined) are the favoured choice for handling finely screened slurries with maximum particle sizes up to about 0.1875 in (5 mm). Where larger solids are concerned rubber linings are liable to damage and hard materials are preferred – *eg* Ni-hard or high chrome irons. A further limitation with rubber-lined pumps is that the head per stage is generally limited to between 90 and 120 ft (30–40 m), whilst Ni-hard or hard fitted pumps can achieve 200 ft (65 m) per stage.

Conventional packed glands are generally suitable for shaft sealing, if water flushed. This type of seal does have certain disadvantages and limitations, however, so individual manufacturers may prefer alternative types, such as the use of expeller vanes, special gland designs or mechanical shaft seals.

PUMPING SEWAGE

Pump types used (see also **Table I**).

(i) Single-channel centrifugal pumps with open (solids handling) impellers and large diameter for handling raw sewage.

(ii) Single- or multi-channel centrifugal pumps for handling treated sewage sludges.

(iii) Mixed flow and axial flow (propeller) pumps for handling large volumes of screened sewage.

(iv) Centrifugal pumps with suitable impellers for handling sewage with gas content, or activated sludge.

(v) Positive displacement (*eg* reciprocating or diaphragm) pumps for handling sludges with low water content and/or where high pressures are required.

(vi) Submersible pumps with open (solids handling) impellers.

(vii) Diaphragm pumps for handling abrasives and/or viscous sludges.

(viii) Plunger pumps for handling heavy viscous sludges.

(ix) Pneumatic ejectors as an alternative to centrifugal pumps for raising sewage to discharge levels.

(x) Packaged sewage pumps – dry sump, wet sump and submersible. In the latter case the lower part of the still forms a wet sump.

Sewage consists mainly of water, a miscellaneous variety of solids in suspension and impurities in solution. Solids can vary considerably in nature and size, and include fibrous materials, grit and miscellaneous waste products as well as fragments of other materials. Untreated or coarse screened sewage has to be lifted from the incoming sewer to a level which will give gravity flow through the primary stages of sewage treatment.

To simplify installation and reduce maintenance pumping stations handling raw sewage are commonly built without screens. The pumps used must then be capable of passing large solids when the only suitable impellers are single-vane or free flow type with spherical passages corresponding approximately to the nominal hose connection. In small pumping stations a minimum of 4 in (100 mm) is commonly specified for the free spherical passage. With larger pumps it is possible to adopt proportionally smaller free passage sections in order to maintain improved pump efficiency. However, the larger the size of solids to be passed by the pump the greater the modification called for in hydraulic design of the unit so that it still remains acceptably efficient in handling finer solids.

A further particular problem in handling raw sewage is that substantial quantities of methane gas may be generated by putrification, resulting in entrained gas as well as dissolved gas being present. In this case the impeller design needs to be of a shape which is largely unaffected by gas and works without congestion. The pumps also need to operate with positive inlet head and with generous intake pipe sizes so that any fall in pressure on the inlet side is kept as small as possible. Similar problems can also occur in sewage treatment plants.

The greater the amount of solids and grit removed by primary treatment the easier it becomes to pump the product efficiently, and the less the wear on the pump. Conventionally Detritus pits are used to remove grit to be deposited and removed by a grit elevator. The sewage is then passed through screens into primary sedimentation tanks where the organic solids settle and form a sludge on

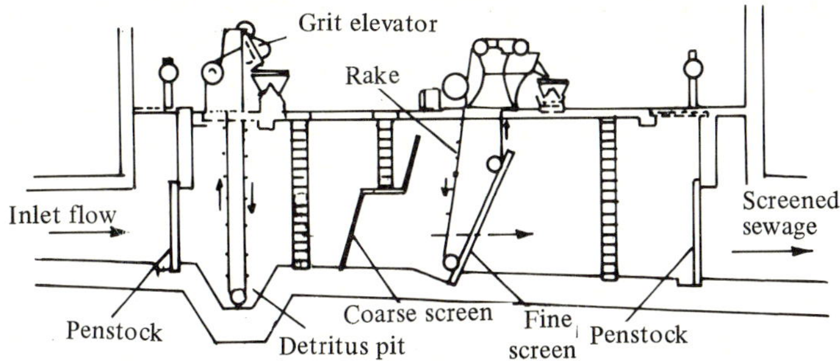

Fig 9.10: A possible arrangement of detritus pit, grit dredger, coarse and fine screens, screen rake, and inlet and outlet penstocks commonly used to remove grit and large solids.

the bottom **(Fig 9.10)**. Grease and fats rise to the surface as a scum. Scum and sludge are then swept to the opposite end of the tanks where they are removed separately.

The sludges resulting from treatment processes can vary from activated sludge and humus sludge with a solids content of 2% or less, to dewatered and digested sludges where the solids content could be as high as 8%.

A problem which can arise in handling sludges with a low water content is the tendency for water to be extracted from the sludge with the possibility of building up a solid 'plug' in the suction head. Modification of centrifugal pump geometry to counter this possibility inevitably leads to a loss of hydraulic performance.

This can be avoided by using a positive displacement pump in conjunction with a centrifugal pump for handling sludges with a low water content. The former pump can then be used to clear any blockage developing.

Stuff Pumps

'Stuff' is the general description for mixtures of water and cellulose pulp which may contain up to 10% of dry pulp (but usually not more than 6%). Centrifugal pumps are the most widely used type for handling such mixtures. Standard impeller forms are suitable for consistencies not greater than about 1.6% when the frictional losses can be anticipated as being similar to that for water alone although the actual pump performance (H-Q characteristics) may be slightly derated. With increasing pulp content it becomes necessary to reduce the number of blades and adopt an open form, both to maintain efficiency and eliminate free spaces in which pulp might collect.

Special forms of impeller may be beneficial, especially when handling high concentrations, since these may be subject to considerably less derating. Thus **Fig 9.11** shows the derating of an Egger impeller handling a 10% concentration compared with that of an ordinary impeller handling somewhat less than one

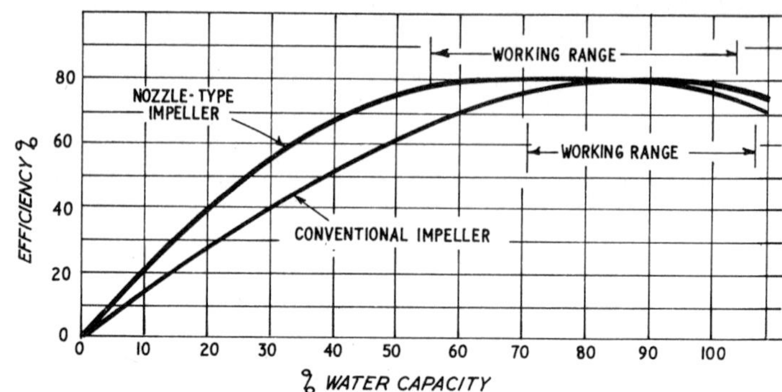

Fig 9.11

half this concentration. A further point with high concentrations (5% or greater) is that it may be beneficial, or necessary, to mount a feed screw leading to the eye of the impeller to ensure positive feed to the pump. Despite the fact that all stuff pumps are invariably operated on flooded suction inflow characteristics can become increasingly irregular with increasing pulp content.

Centrifugal stuff pumps may still be rated on water performance since this forms a consistent basis for correction, as well as being the easiest method of establishing test figures since only one liquid is involved. Head-capacity corrections may then be determined for the type of stock and stock consistency, although this can only be expected to give approximate rather than exact solutions.

Typical head-capacity correction factors for chemical stock are given in **Fig 9.12**; and those for mechanical and reclaimed stock in **Fig 9.13**. The appropriate factor is applied both to head and capacity thus –

(i) suppose the capacity required is Q gal/min

(ii) suppose the total head required is H ft

Find the appropriate factor (f) for the consistency and type of stock involved. Then, based on the water rating figures the pump required should –

(iii) deliver a capacity of Q/f at a head of H/f

The head-capacity factor may need further correction for the nature of the stock. Thus lower factors may apply where the stock contains entrained air or is freer and less readily maintained in suspension. Other types of stock, particularly those with fillers and additives, may be more homogeneous and requiring less correction (*ie* a higher correction factor), unless these induce gassing, when the opposite effect may be apparent. In cases of doubt it is usually possible to establish a suitable correction factor by empirical tests on a suitable pump, when this factor can be applied with a reasonable degree of confidence to other pumps of similar type. In practice it is generally satisfactory to select a pump size where

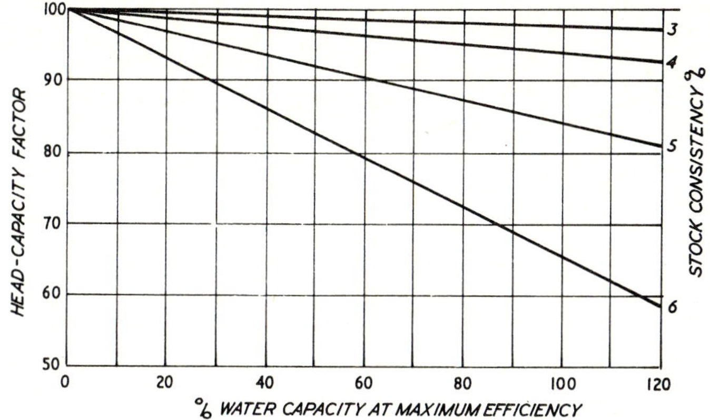

Fig 9.12: Correction factor for chemical stock.

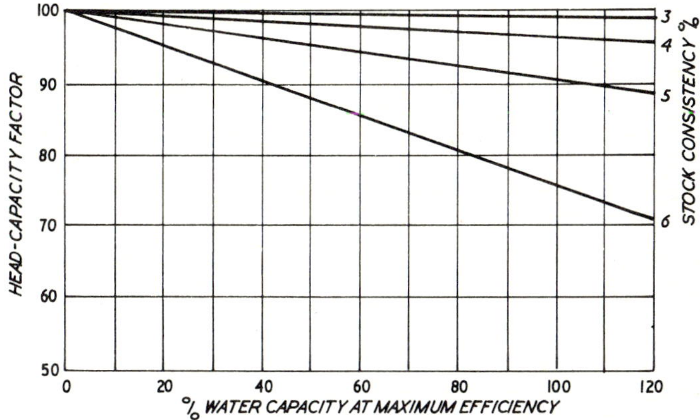

Fig 9.13: Correction factor for mechanical and reclaimed stock.

the water H-Q working point comes within the working range defined by the pump envelope on its rating chart. In the case of selecting the most suitable impeller diameter for a particular unit several trial calculations may have to be made to arrive at an impeller size giving maximum efficiency.

Other Pump Types

Where other types of pumps are used to handle stuff the head-capacity correction factors discussed above do not apply. With positive displacement pumps capacity is little affected by head and is mainly related to the speed of the pump for a given size. Derating of performance may, however, apply since it may be necessary to reduce the operating speed to ensure consistent filling of the pump.

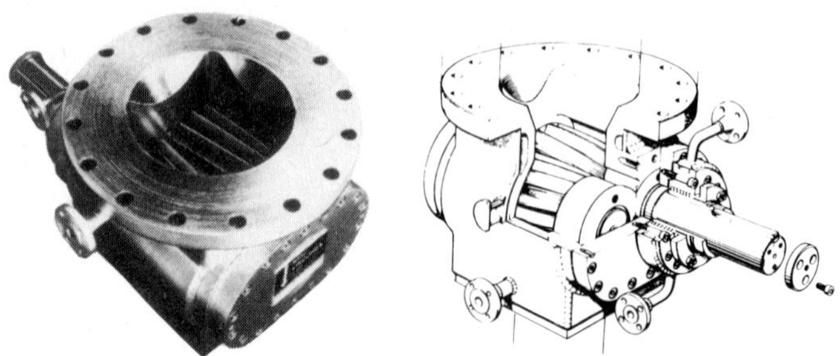

Maag jacketed gear pump for handling high viscosity fluids at temperatures up to 300°C.

In this case similar considerations apply as for viscous fluids (which see), although the pump type chosen must be suitable for passing the suspended material without jamming or clogging; or undue loss of efficiency resulting from the employment of generous clearances.

Frictional Losses

Frictional losses through the delivery pipeline are significant in handling stuff with a consistency of 2% or greater since the friction head (and thus total head) will increase rapidly with increasing consistency. The friction factor is dependent on the type of pulp and a pseudo-Reynold's Number, *viz*

$$\text{friction factor (f)} = \frac{K^1}{R_e^1}$$

where:

K^1 is a constant for a particular type of pulp

R_e^1 = pseudo-Reynold's Number

$$= \frac{17.2\,Q}{C^{1.57}\,D^{1.795}}$$

where:

Q is in gal/min

C = stock consistency %

x = an exponential = 1.63

D = pipe diameter in feet

This formula is derived empirically, but is extremely tedious to work. It is more usual to refer to available data published in the form of friction loss against flow for individual pipe sizes – *eg* **Fig 9.14** shows typical values for a 6 in (150 mm) pipe. The slope of the curves remains the same for all diameter sizes

FRICTION LOSS WITH PULP STOCK—12 in. PIPE

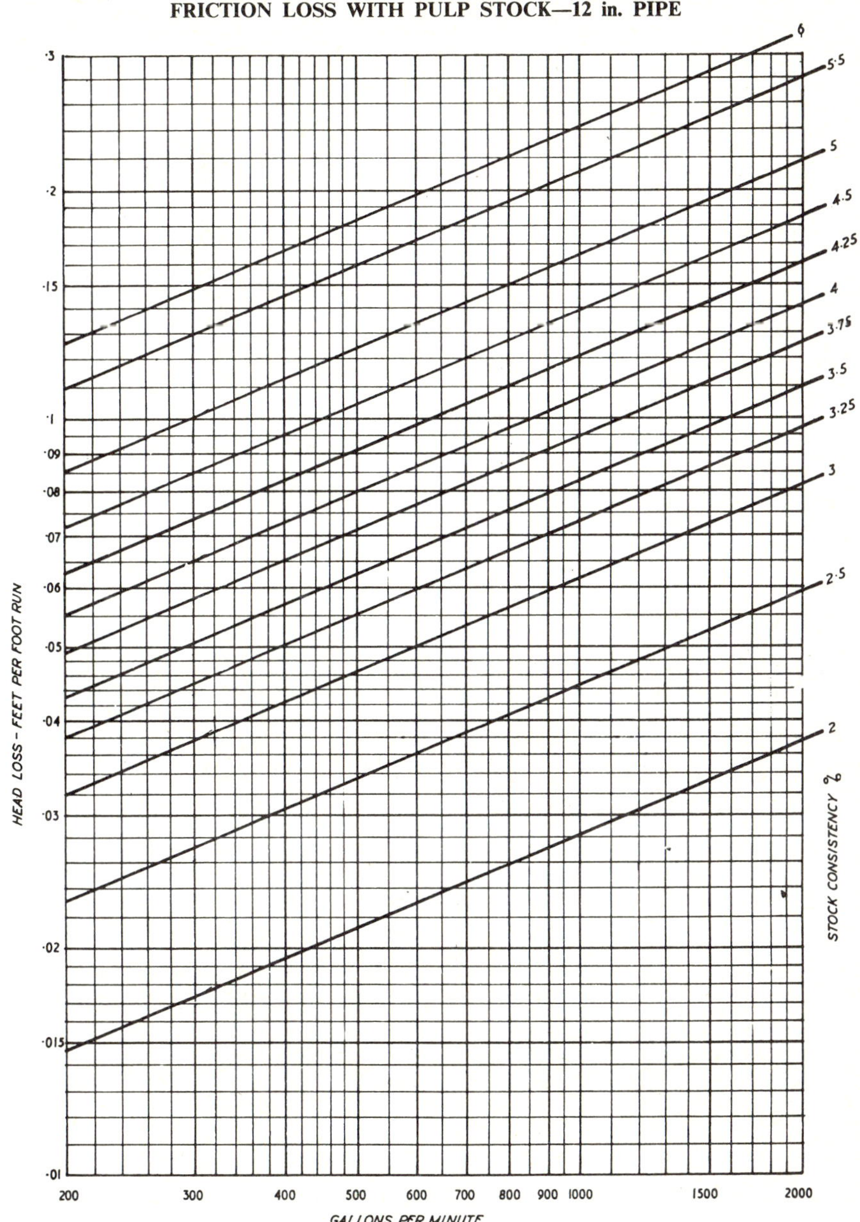

Fig 9.14

and lacking further data this particular curve can be corrected for other diameter sizes on the basis

$$\text{Friction loss} \quad \propto \frac{1}{(D^{1.33})\ (D^{2.37})}$$

$$\textit{ie}\ \text{approx} \quad \propto \frac{1}{D^4}$$

Also frictional loss is proportional to $Q^{0.37}$; or approximately proportional to $\sqrt[3]{Q}$.

Strictly speaking these data are applicable only to the type of stuff evaluated (unbleached sulphite) and may need correction for other types of pulp. Typical correction factors involved are –

0.9 for bleached sulphite, soda, sulphate and reclaimed paper stocks.

1.2 to 1.4 for groundwood.

Two CD60C3/5 Mono pumps feeding paper stock consistency controllers at the rate of 10 gal/min handling stock at 3.5–6% consistency.

Homogeneous Fluids

Fine solid particles in suspension in a liquid can form a homogeneous mixture, provided the flow is turbulent. If flow is laminar, settling out of particles is likely and the mixture is no longer homogeneous.

Mixtures which are truly homogeneous can be classified as real liquids, with flow properties similar to that of a simple fluid of the same viscosity (although the viscosity of the homogeneous fluids is an apparent one). Normal corrections can then be applied to water performance as for viscous fluids, particularly for the calculation of pipeline losses. If, however, the concentration of solids is too high the mixture will have no true (apparent) viscosity and the homogeneous fluid has non-Newtonian characteristics. It cannot then be treated as a simple fluid. Pastes and gels also cannot be regarded as true fluids, although they may be homogeneous mixtures.

Sensitive Fluids

Certain classes of fluids can only be handled with a gentle pumping action since they are sensitive to shear or agitation which could produce a permanent loss of properties. It becomes essential with such fluids to choose a pump which does not generate high local shear on the fluid and/or produce strong mechanical agitation. Choice is thus virtually limited to positive displacement types capable of producing steady (non-pulsating) flow in an axial direction.

The problem can normally be resolved by:

(i) Selecting a pump type with a suitable gentle action.

(ii) Determining the maximum speed for the pump which can be tolerated by the product by actual test.

(iii) Selecting the size of pump on the basis of maximum speed determined in (ii) and the capacity required.

Capacity is not necessarily of primary importance in such cases and so some sacrifice of delivery may be fully acceptable. The main point here is that the available capacity consistent with the acceptable maximum speed may be very much lower than the nominal capacity of the pump chosen, because the speed may have to be reduced considerably. Also a suitable pump type and size may have quite different limiting speeds for handling non-Newtonian fluids of similar 'family' characteristics but different sensitivities to mechanical agitation.

10. Materials and Compatability

GENERAL PURPOSE centrifugal pumps, handling water, solvents and caustic solutions commonly have casings made in cast iron with impellers in cast iron or bronze. For handling mildly acidic fluids, bronze or stainless steel is a normal choice. Where more aggressive acids are to be handled, acid-resisting stainless steels, alloy steels, silicon cast iron, or even titanium may be employed; alternatively a lined pump or a pump with non-metallic wetted parts (*eg* one of the modern engineering plastics which show exceptional corrosion resistance.

With other types of pumps it is becoming more usual to employ materials with a wide range of compatibility, with alternative materials available, if necessary, to deal with special problems of chemical attack or erosion-corrosion. High duty materials tend to be expensive (and in many cases present fabrication problems), so there is also an increasing tendency to replace the more expensive metals with cheaper materials, such as elastomers and plastics, where the size, design and service temperature permit. This has been further encouraged by the continual development of man-made materials of improved physical and chemical properties, and a reduction in cost of such materials through quantity production.

A broad guide to material selection is given in **Table I**. A general classification on this basis, however, has distinct limitations, since it can embrace a range of alloys which can vary considerably in corrosion resistance. Thus stainless steel is a broad, rather than a specific, material description which can include a whole range of nickel-chromium alloy steels. Chemical resistance (and cost) increases with increasing alloy content and can be further enhanced by small additions of molybdenum or tungsten or cobalt. The properties of austenitic stainless steels can also be modified by heat treatment or working in a hot state (*eg* welding), as this can cause carbon precipitation and intergranular corrosion. This limitation can be overcome by the introduction of stabilising elements, such as columbium or titanium.

As far as metals are concerned, in fact, the pH value of the product wetting it is an unreliable guide. What matters, if the liquid is known to be chemically active, corrosive or otherwise aggressive is the specific resistance of particular metals to that product, at the temperature and under the conditions at which it will be handled.

TABLE I – PUMP MATERIALS RELATED TO pH

Acidic		pH Neutral	Basic	
0–4	4–6	7	8–9	9–14
Stainless steels **Acid-resisting steels** Alloy steels Silicon cast iron Titanium Carbon Ebonite Ceramic (lined) Rubber (lined) Plastic (lined) Non-metals*	**Bronzes** Non-metals*	**Cast iron** **Cast iron (bronze fitted)** Cast steel Aluminium Non-metals*	**Cast iron** Non-metals*	**Cast iron** Nickel Nickel alloys Non-metals*

Metals in **bold** are 'traditional' or standard choices, but note the term 'standard construction' normally applies to cast iron construction for series with pH 6–9.

*Non-metals is descriptive of modern engineering plastics.

The onus, basically, is on the pump manufacturer to provide a suitable material, although such a choice must be analysed on the basis of service experience if possible; and if the choice is an exotic alloy, the cost justified. Titanium, for example, is an experience-proven metal suitable for most chemical pump constructions, but extremely expensive and difficult to fabricate (although it is readily castable). Its cost could be justified by long service life.

Increasing use is being made of non-metallic materials for chemical pump constructions. Whilst these materials have lower mechanical properties than metals (and higher moduli which can present problems of differential expansion/contraction in mixed constructions), many offer exceptional chemical resistance, even extending over the full pH range of 0–14. With non-metals, too, corrosion resistance expressed in terms of pH range *is* a valid guide and the parameter normally employed in assessing compatibility.

Solving the Problem

The problem of satisfactory material selection can be summarised as follows:

(i) For non-critical applications, or where the fluid has substantially neutral characteristics (pH 6–9), standard constructions should be generally satisfactory and less costly than the use of special materials.

(ii) For applications involving fluids which are more acidic (pH 6) or more basic (pH 9), general recommendations may not be fully satisfactory and the question of the best material to employ may need more detailed, or even specific, analysis. To some extent this is bound up with the type of pump selected, some types being more readily adaptable to construction in alternative materials than others.

Compatibility

Compatibility, or corrosion resistance as established by laboratory tests is not enough. The actual performance of a particular metal in contact with a particular fluid under working conditions may differ appreciably from the results obtained by static tests. The chief causes of such differences are:

(i) Catalytic effects – the presence of contaminants in the fluid can often produce catalytic action accelerating corrosion.

(ii) Erosion-corrosion – metals which are resistant to chemical attack by virtue of the building up of a passivated surface layer may have that layer continually worn away under working conditions and thus remain open to attack. This is more usually confined to localised areas in a system subject to high local flow velocities. Erosion-corrosion can also occur if the fluid contains abrasive particles, or even with a clean fluid containing entrapped air.

(iii) Electrolytic corrosion – this can occur at contact joints between dissimilar metals, the use of weld metal with incomplete compatibility and also at discontinuities in a homogeneous assembly. Electrolytic attack is also enhanced by the presence of aeration in the fluid. Electrolytic-corrosion and erosion-corrosion are often closely linked or combined.

(iv) Stress-corrosion – this can occur at discontinuities in a nominally homogeneous metal resulting in improper working of the material and intercrystalline cracking. In some cases cracking may be the result of working stresses.

The above descriptions are somewhat over-simplified, but serve to illustrate that the question of complete compatibility can be a most complex and demanding one. The problem also tends to become aggravated by unusual working conditions, such as high temperatures. It is thus only possible to establish completely reliable data under actual working conditions identical to that of the system under consideration. Close co-operation between user and pump manufacturer is therefore necessary to utilise the latter's experience in the field concerned. Further comment on material suitability will also be found in Chaper 7, where appropriate.

Centrifugal Pumps

Cast iron remains a standard choice for castings for all sizes of general purpose pumps, normally employing modern close grain SG irons. Cast steel may be used where higher pressures and temperatures are involved particularly for (i) smaller casings and (ii) high speed impellers.

Casings are normally cast, although in special constructions they may be machined from solid. Impellers are normally cast also, although in special constructions they may be fabricated by welding; and where plastic materials are used they may be produced by moulding. The application of cast iron for impellers is nominally limited by size and peripheral speed. Smaller impellers are usually cast in bronze because of the small cross sections involved, whilst larger bronze castings are also used for all-bronze or bronze-fitted pumps. Shafts are invariably of high tensile steel (with stainless steel usual for special constructions) and seals are of the packed gland type, although mechanical seals may be offered as an alternative.

Cast iron is generally suitable for handling all types of solutions ranging from neutral (*eg* water) to the most basic (pH = 14); although monel may be preferred in special circumstances for handling strong alkalis, or plastics for reduced cost. Pumps for handling slight to moderately acidic fluids are normally of bronze of varying compositions, or aluminium may be preferred (*eg* for milk).

Non-metals (other than plastics) include (chiefly) carbon and ebonite (hard rubber). These have rather limited application as structural materials and are more generally used as linings, see **Table II**. They are classed separately to ceramic materials since the latter are invariably used as linings, or with composite construction, because of their brittle nature and the difficulty of working them. Materials in this category include porcelain, stoneware and glass, usually cemented into a housing of cast iron.

Centrifugal pump impellers in polyvinylidene fluoride (Kynar)

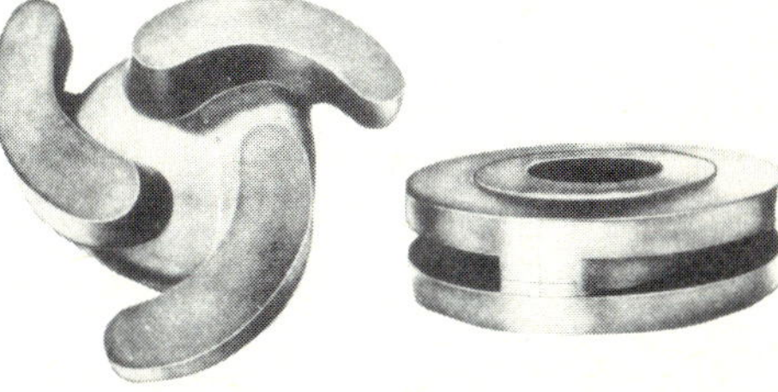

Open and closed rubber covered impellers. (Allis Chalmers)

TABLE II – GENERAL GUIDE TO CORROSION RESISTANCE OF STAINLESS STEELS

LIQUID	Stainless Steel Type			
	Austenitic AISI 302 (18/8)	Austenitic AISI 316 (18/8 med)	Ferritic AISI 430	Martensitic AISI 410
Acids: generally suitable except for the following:				
Acetic acid	L	L	L	L
Arsenic	L			
Boric	L	S	L	
Carbolic	L	L	X	X
Chromic	X	X	X	X
Formic	X	L		X
Oxalic	L	L	X	X
Sulphuric	L	L	X	X
Tartaric	L	L	L	L
Aqueous solutions: generally suitable except for the following:				
Alum	L	L		
Ammonium Chloride	L	L	L	L
Bleaching Powder	L	L	X	X
Calcium Chloride	L	L	X	X
Calcium Hypochlorite	X	L	X	X
Copper Chloride	X			S
Copper Sulphate	S	S	S	L
Hydrogen Peroxide	L	L	L	
Magnesium Chloride	L	L	X	X
Potassium Chloride	X	L	X	X
Sodium Chloride	L	L		
Sodium Hypochlotite	L to S	L to S	X	L
Sodium Thiosulphate	L	L	L	X
Stannic Chloride	X	X		X
Stannous Chloride	L		X	X
Zinc Chloride	X	L		X
Solvents: generally suitable except for:				
Carbon Tetrachloride	L	L	L	L
Trichlorethylene	L	L	L	L
Organic Chemicals: generally suitable except for:				
Formaldehyde	L	L	L	L
Misc. fluids: generally suitable except for:				
Glues	L	L	L	L
Inks	L	L	L	L
Bromine (& Bromine Water)	X	X	X	X
Chlorine (wet or dry)	X	X	X	X
Hydrogen Sulphide	L	L		
Lysol	L	L	X	X
Sulphur Dioxide	L	L		
Foodstuffs: generally suitable except for:				
Tomato Juice	L	L	L	L
Vinegar	L	L	L	L
Waters: generally suitable except for:				
Mine Water	L	L	L	L
Sea Water	L	L		X

(S – Suitable L – Limited Suitability X – Unsuitable)

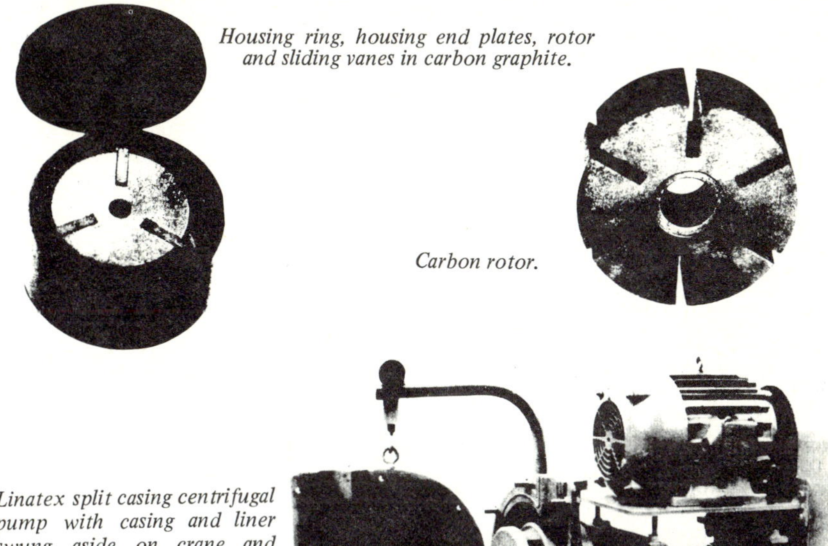

Housing ring, housing end plates, rotor and sliding vanes in carbon graphite.

Carbon rotor.

Linatex split casing centrifugal pump with casing and liner swung aside on crane and crawl. The pump has only three wearing parts: Gland side liner, Suction side liner, and Impeller.

A much wider variety of materials are employed for handling the more acidic fluids. These can be in the form of structural materials, or linings or coatings (applied to casings and impellers, respectively). The materials used can be classified as (i) metals, (ii) non-metals, (iii) ceramics and (iv) plastics. Each class has its respective advantages and disadvantages.

Metals used include stainless steels; non-ferrous alloys based on nickel and chromium or molybdenum with less than 20% iron; high-silicon iron with a minimum silicon content of 14.25%; austenitic cast iron with a minimum of 22% of nickel, copper and chromium; aluminium; lead hardened with antimony, and a variety of proprietary high-duty alloys. Chief advantages offered are specific and consistent physical properties and high chemical resistance (although this may be related to specific alloys). Chief disadvantages are high cost in many cases, both as regards the material itself and the fabrication techniques which may be involved.

Lead, rubber and glass were originally used as linings or coatings. Rubber is still used with a wide range of synthetic elastomers available; and glass still has application for hygienic services (and is particularly suitable for sterilisation).

Plastic materials are now normally favoured as providing a greater resistance to a wider variety of media than rubber-lined, or all-metal pumps. These include PVC, polythene (PE), polypropylene (PP), polyvinylidene fluoride (PVDF) and PTFE. Additionally, copolymers of these materials may be used – see also **Tables III** and **IV**.

TABLE III – SOME MODERN NON-METALLICS COMPARED

Plastic	Properties
High molecular PE	Compares favourably with rubber-lined or hard metal lined pumps. Withstands almost all acids and alkalis up to 90°C (conc nitric acid excepted). Resistant to most (but not all) solvents.
PP	Similar chemical resistance to PE, but resistance to aggressive media not quite as good.
PDF	Exceptional resistance to most media, especially chlorinated fluids. Very high maximum service temperature for a plastic, *ie* 120–130°C.
PTFE	Resistant over the full pH range 0–14 with a maximum temperature resistance of 150°C in pump applications. Relatively low mechanical strength and resistance to abrasion and wear rather poor.

TABLE IV – RESISTANCE TO ACIDS OF MODERN PLASTICS

	Hydrochloric Acid	Nitric Acid	Sulphuric Acid	Phosphoric Acid	Sodium Hydroxide
Polyvinylidene Fluoride	E	G	G	E	G
Polytetrafluoro-ethylene (PTFE)	E	E	E	E	E
Epoxy	E	F	F	G	E
Polypropylene	G	F	G	E	E
Chlorosulphonated Polyethylene	G	F	G	F	E
Vinyl Ester	E	F	G	E	G
Polyvinylchloride (PVC)	G	P	F	E	F
Glassed Steel	E	E	E	E	E

E = Excellent G = Good F = Fair P = Poor

The development of high strength plastic materials has made it possible to utilise solid plastic components in pumps such as nylon or polyacetal impellers (and even casings in smaller sizes). In general, however, for anything other than small pump sizes, reinforced plastics are necessary in order to achieve the required mechanical properties. Particular developments have been made in this respect with glass reinforced plastics based on thermoset resins as being considerably superior in mechanical strength and abrasion resistance.

Glass-fibre reinforced materials based on polyester resins can be fabricated into almost any form and are therefore especially valuable for certain sizes and designs of equipment. With a temperature limit of approximately 110°C they are particularly resistant to salt solutions and to any fluids acidified with hydrochloric acid, though their application is much wider than that.

Epoxy resin based GRP is an even more versatile reinforced plastic and in pump applications may be reinforced with carbon/graphite fibres as well as fibreglass. In Wernert pumps for example this mixture is known as Durapox, and a temperature limit of 125°C is permissible. It is particularly useful for its resistance to solvents, to H_2SO_4 and to a wider range of media than the polyester based GRP.

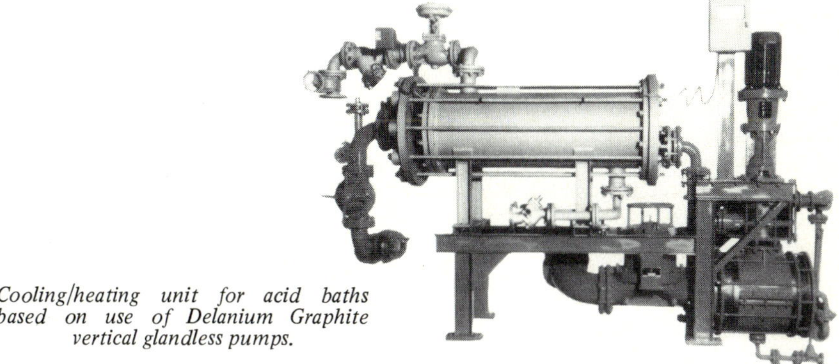

Cooling/heating unit for acid baths based on use of Delanium Graphite vertical glandless pumps.

Other Types of Pumps

Choice of constructional materials follows on similar lines, except that the form of the pump may considerably reduce the number of alternatives or special constructions necessary. Thus it may be practical to produce a particular design in stainless steel as standard for both general purpose or chemical duties without suffering a noticeable cost penalty; or reduce cost to a minimum on a quantity produced item by employing a maximum number of components injection moulded in plastic.

Generalisation is difficult because of the different forms and duties of the individual components involved in different pump designs, and also the different stresses they may have to carry. The extension in the use of plastic materials where they can be utilised, however, is a rational one since the chemical inertness of suitable selected plastics extends the application of the pump to the handling of a wider range of fluids without modification, although the main application is to internal components or wetted surfaces (except in the case of very small pumps with low component stresses which can be all-plastic except for the shaft).

Metals are generally to be preferred for all casings and all components where high dimensional rigidity is required, whether stressed mechanically or thermally. All plastic materials suffer from much higher modulii than metals, are more temperature sensitive and have substantially lower maximum surface temperatures. Thus, normally, any conventional type of pump would be constructed in suitable low-cost metals for general purpose duties where corrosion is not a problem in order to provide maximum rigidity and durability at minimum cost. Alternative materials of construction may then be offered to extend the range of application to other fluids. This is particularly the case with nearly all medium and large size machines; although to avoid the complexity of too many alternatives in special constructions, some manufacturers of smaller machines may prefer to adopt more exotic materials initially, or as a single alternative to provide for multiple duties.

It is the manufacturer, rather than the user, who remains primarily responsible for material selection and can then, for a specific pump and specific materials –

(i) State broad compatibility – *ie* suitability for different types of fluids.

(ii) Estimate specific compatibility – *ie* given the actual fluid to be handled, working temperature and all appropriate information governing working conditions, recommend the most suitable materials from his range.

(iii) Specify any limitations – *eg* notably temperature limitations.

Item (i) is generally covered by manufacturers literature and general specifications. Item (ii) may also be forthcoming if the manufacturer specialises in the production of multi-duty pumps, but such data may vary from comprehensive, to very incomplete. If the particular fluid concerned is not covered, or the working conditions are likely to be unusual, co-operation with the pump manufacturer is necessary to investigate the problem further. Item (iii) may very well be covered by (ii) and is usually readily available in any case where non-metallic materials are concerned.

Ideally the problem of materials and compatibility should be presented to the manufacturer, not solved by the user, as the former's experience in this field is likely to be the greater. Also the user is normally concerned with handling a fluid and not material properties as such, which are quite a different study. However, the user may require some basis or background knowledge to assess recommendations from different sources – and which in some cases may appear conflicting. The various tables included in this chapter have been drawn up with this in mind and should be useful for both general and specific reference, although they must not be read as completely authoritative.

11. Pipework Calculations

PIPEWORK calculations are concerned with (i) determining the frictional head losses (or pressure drop) in the pipework system; and (ii) determining suitable pipe sizes. Each represents a separate series of calculations or estimates which may set conflicting requirements – *eg* a large pipe diameter will produce minimum friction for a given flow rate but will also represent high initial cost. Thus the main object is (i) to determine frictional losses which are a necessary part of establishing the total system head on which the selection of a suitable pump is based; and (ii) finally determine an optimum size of piping. It should be noted that other factors pertinent to *system design* may be involved, but these will fall into the two categories mentioned.

Flow Losses

In a good many cases flow losses are estimated on empirical or semi-empirical lines, or even 'guesstimated'. Alternatively, tables or charts may be used to read friction losses direct for particular flow rates and pipe sizes. The accuracy obtained depends very much on the validity of the data employed, many of which show remarkable inconsistencies or divergencies. The majority of such data, too, are prepared for water flow and are thus not directly applicable to fluids with higher viscosity. Thus, unless tabular or other data available are known to be reliable, and are applicable to the liquid being handled, more accurate results can usually be obtained by direct calculation.

Type of Flow

The type of flow, whether *laminar* or *turbulent,* can be determined from the flow Reynolds Number (R_e). This is a quantity expressed by dividing the product of pipe bore and flow velocity by the kinematic viscosity of the fluid involved. Although a dimensionless quantity, the actual numerical value of R_e obtained is dependent on the units employed for bore and flow velocity (assuming that viscosity ν is always quoted in centistokes).

In engineering units:

$$R_e = \frac{7740DV}{\nu}$$ where D is in inches, V is in ft/sec

$$R_e = \frac{930DV}{\nu}$$ where D is in cm, V is in m/sec

TABLE I – REYNOLD'S NUMBER FOR CLEAN COLD WATER**

Inches (mm)	PIPE BORE									
	1 (25)	1½ (40)	2 (50)	3 (75)	4 (100)	6 (150)	8 (200)	10 (250)	12 (300)	18 (450)
Per 1 gal/min*	3800	2500	1900	1270	950	630	475	380	320	210
Per 1 lit/min*	835	550	420	280	210	140	105	85	70	46

* Multiply by actual numerical value in either unit to give Reynold's Number

** Extracted from Handbook of Valves, Piping and Pipelines

Note: To determine R_e for any other fluid divide value for water by fluid viscosity in centistokes.

For clean, cold water

$R_e = 7740\ DV$ or $R_e = 930\,DV$

(see also **Table I**)

Laminar Flow Losses

Where the fluid flow is *laminar* (*ie* Reynold's Number below 2 000, (or 240 metric), accurate calculation of frictional losses can be made from first principles using the Darcy-Weisbach formula in the form

$$\Delta P = f \cdot \frac{L\rho V^2}{2Dg}$$

where:

P = pressure drop

ρ = mass density of fluid

f = a friction factor = $\dfrac{64}{R_e\ (\text{English})}$

Rendered in engineering units, and rewritten to include flow rate (Q) rather than flow velocity this becomes:

$$\Delta P = K_1 \frac{LQ\mu}{D^4}$$

where:

L is the length of the pipe

μ is the viscosity in *centipoise*

K_1 is a factor depending on units for L, Q and D.

For L in feet and D in inches:

$K_1 = 0.0002275$ for Q in Imp gal/min

$= 0.000273$ for Q in US gal/min

For L in metres, D in millimetres and Q in litres/min

$K_1 = 9.05$

Expressed as a head loss (ΔH), the equivalent engineering formula is:

$$\Delta H = K_2 f \cdot \frac{LQ^2}{D^5}$$

For ΔH and L in feet and D in inches:

$K_2 = 0.026$ for Q in Imp gal/min

$= 0.0311$ for Q in US gal/min

For ΔH and L in metres, D in millimetres and Q in litres/min

$K_2 = 641\ 270$

Note that neither formula includes the Reynold's Number as an individual factor (although it is effectively combined in the equation). Thus it is necessary to calculate the Reynold's Number of flow separately to determine whether or not the flow is laminar and the formula can be used. A 'short cut' method here is to calculate the quantity $\nu D/Q$ (viscosity in centistokes x pipe bore in inches, divided by flow rate). If this is greater than 0.4 with the flow rate in m^3/sec, or greater than 1.85 with the flow rate in gal/min, the flow will be laminar.

Note also that the laminar flow formula applies regardless of the condition of the pipe bore, *ie* whether smooth or rough, and is thus directly applicable to laminar flow in all types of circular pipes. Unfortunately in a majority of practical cases of pumped fluids the Reynold's Number of flow is turbulent and simple calculation of flow losses on the above basis is no longer possible.

Turbulent Flow Losses

In the case of turbulent flow the friction factor is inversely proportional to an exponential value of the Reynold's Number and also to the surface condition or surface roughness of the pipe. The former (exponential) can be derived semi-empirically, but the latter can only be established by empirical tests. It is more convenient to combine the two in the form of a single friction factor related both to Reynold's Number and *relative roughness,* **(Fig 11.1)**.

Relative roughness is a non-dimensional quantity defined as the effective height (ϵ) of the projections formed by the bore roughness divided by the pipe bore diameter (D), *ie*

relative roughness $= \epsilon/D$

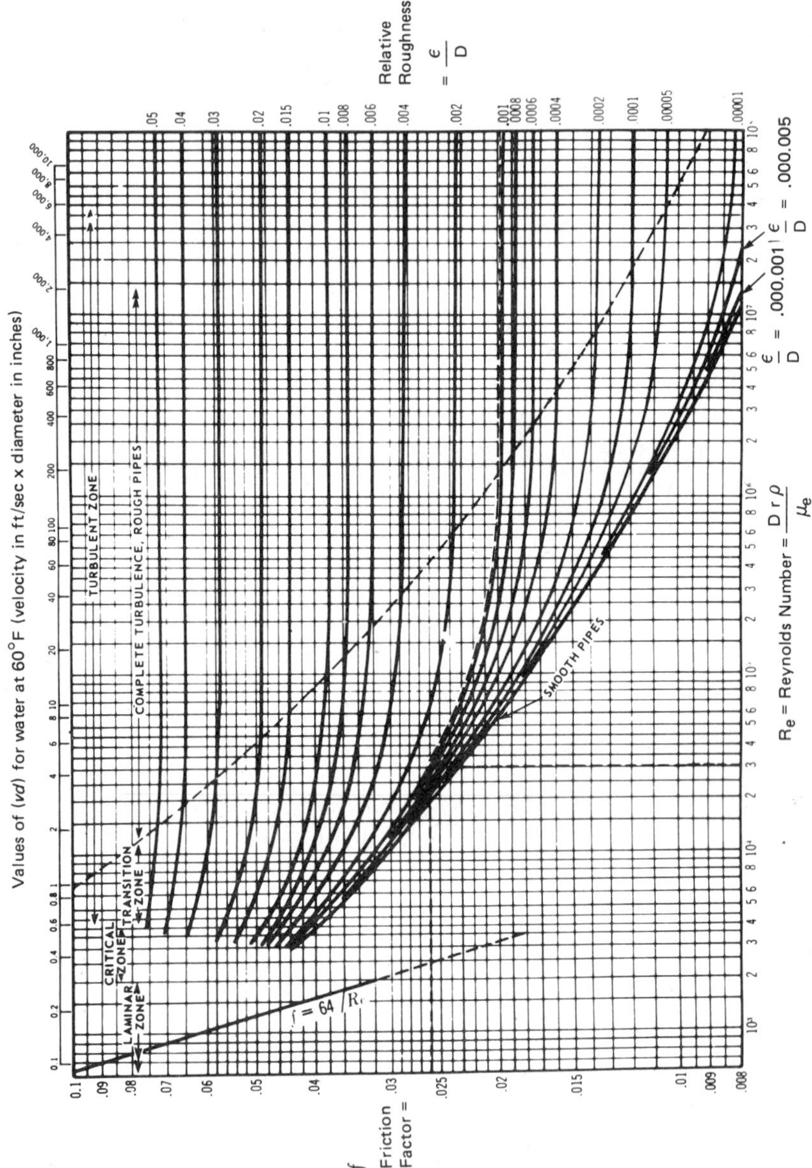

Fig 11.1: Friction factors for pipes.
Example: friction factor for pipe with relative roughness 0.001 at flow Reynolds Number of 30 000 = 0.026.

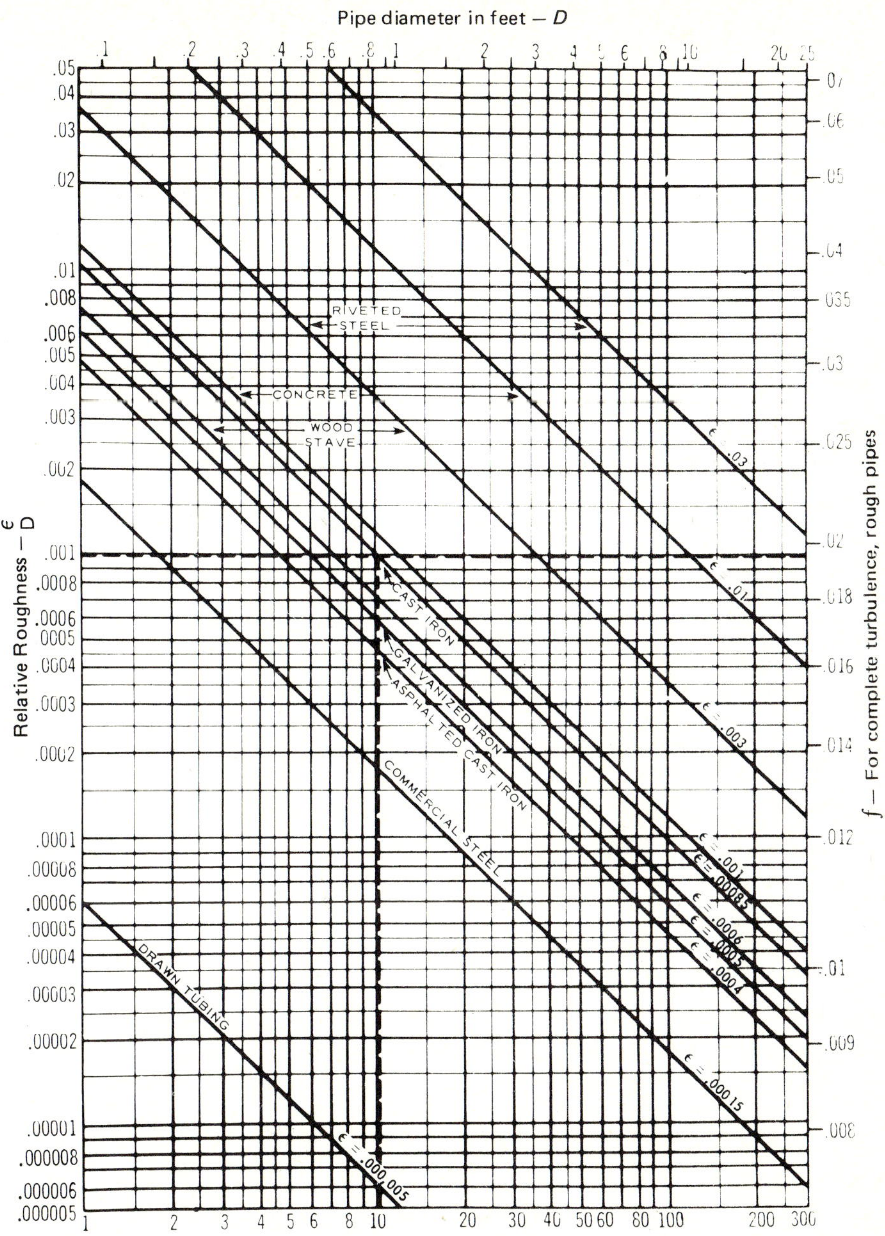

Fig 11.2: Friction factors and relative roughness for commercial pipes with fully turbulent flow (based on ASME data originated by L. F. Moody). Example, for 10 in diameter cast iron pipe, relative roughness (E/D) = 0.00085. Friction factor = 0.0196.

TABLE II – SURFACE ROUGHNESS (K) FOR PIPES

Pipe Material	New Pipes K (in)	New Pipes K (mm)	Old Pipes K (in)	Old Pipes K (mm)
Plastic	0.0004	0.01	0.01	0.25
Drawn steel	0.002	0.05	0.04	1.0
Welded steel	0.004	0.10	0.04	1.0
Drawn stainless steel	0.002	0.05	0.01	0.25
Welded stainless steel	0.004	0.10	0.01	0.25
Cast iron	0.01	0.25	–	–
Galvanised steel	0.006	0.15	–	–
Concrete	0.01–0.08	0.3–2.0	–	–
Asbestos-cement	0.001	0.025	–	–

It is impractical to measure k directly and so relative roughness is normally estimated by reference to typical values – see **Fig 11.2** and **Table II.** The friction factor can then be read directly from the graph shown in **Fig 11.1**, against the appropriate Reynold's Number. This graph also shows the four phases of flow condition – (i) laminar flow up to RN = 2 000; followed by (ii) a transition period where the flow condition is unstable and unpredictable; (iii) a region of developed turbulent flow where the friction factor decreases with increasing Reynold's Number as well as increasing with increasing roughness; and finally (iv) the region of fully developed turbulent flow where the friction factor is a constant regardless of increase of Reynold's Number and dependent only on the surface roughness.

Note: Charts of similar form to **Figs 11.1** and **11.2** may show substantially different values of friction factor for similar values of relative roughness. This is because the general formula is of the form $\Delta H/L = f \cdot \dfrac{Q^2}{kD^5}$ which includes the Reynold's Number as a factor. The value of the friction factor (f) derived is thus adjusted accordingly.

Working formulas are:

$$\frac{\Delta P}{L} = K_1 f \cdot \frac{\rho V^2}{D}$$

For ΔP in lb/in^2, ρ in lb/ft^3, V in ft/sec and D in inches, $K_1 = 0.001294$

For ΔP in bar, ρ in tonnes/m^3, V in metres/sec and D in millimetres

$K_1 = 0.000001125$

$$\frac{\Delta H}{L} = K_2 \cdot \frac{V^2}{D}$$

For ΔH in feet, V in ft/sec and D in inches, $K_2 = 0.1863$
For ΔH in metres, V in metres/sec and D in millimetres, $K_2 = 0.041$.

It is generally more convenient to work these formulas in terms of flow rate Q, *viz:*

$$\frac{\Delta P}{L} = K_3 \, f \cdot \frac{\rho Q^2}{D^5}$$

For ΔP in lb/in^2, ρ in lb/ft^3, D in inches – $K_3 = 0.00018$ for Q in (Imp) gal/min; or 0.000216 for Q in (US) gal/min.
For ΔP in bar, ρ in $tonnes/m^3$, Q in lit/min and D in millimetres, $K_3 = 0.1613$.

$$\frac{\Delta H}{L} = K_4 \cdot f \cdot \frac{Q^2}{D^5}$$

For ΔH in feet, D in inches – $K_4 = 0.026$ for Q in (Imp) gal/min; or 0.0311 for Q in (US) gal/min.
For ΔH in bar, Q in lit/min and D in millimetres, $K_4 = 641\,270$.

Losses in Bends and Fittings

Evaluation of frictional losses by the method described above is applicable only to straight runs of piping of constant diameter. Additional losses will be involved at bends and fittings, and changes in section.

Losses in bends and fittings cannot be calculated directly but must refer to empirical data. There are basically two methods of analysis, *viz:*

(i) The fitting resistance is expressed in terms of the *equivalent straight length* of pipe (*ie* the length of straight pipe of the same diameter as the fitting which would have the same resistance). This is the simplest method as the total resistance of the fittings in the circuit is then expressed in terms of an additional pipe length for calculation of total pipework resistance. Typical data are given in **Table III.**

(ii) Fitting resistance is given in terms of a *resistance coefficient* (K) when losses can be calculated directly from the formula:

$$\Delta P = \frac{KV^2 w}{2g}$$

$$\text{or } \Delta H = \frac{KV^2}{2g}$$

TABLE III – EQUIVALENT STRAIGHT LENGTHS
(Note: Equivalent Length = Table Factor x Pipe Diameter in same length unit as diameter).

Flush sharp-edged entry	20–25	Branch piece, flow to branch	45
Slightly rounded entry	10–12	Branch piece, flow from branch	22
Flush bellmouth entry	4	Bellmouth outlet	9
Sharp entry projecting into liquid	35	Sudden enlargement	45
Bellmouth entry projecting into liquid	10	Taper, divergence angle above 60°	45
Footvalve with strainer	100	Taper, divergence angle 15–60°	22
Round elbow	45	15° divergence angle	1
Short radius bend	30–35	Screwdown valve, straight	100
Medium radius bend	15	Screwdown valve, angled	50
Close return bend	100	Gate valve	7
Tee: straight through	10	Gate valve, half closed	15
side outlet, sharp angled	50–55	Gate valve, three-quarters closed	60–65
side outlet, radiused (swept tee)	20–25	Globe valve	250–450
Branch piece, straight through	10	Swing check valve	60–70
		Reflux (non-return – valve)	45

For clean, cold water:

$$\Delta P\ (\text{lb/in}^2) = 0.00673\ KV^2 \text{ (where V is in ft/sec)}$$
$$\Delta P\ (\text{bar}) = 0.000044\ KV^2 \text{ (where V is in metres/sec).}$$

The value of the resistance coefficient (K) is independent of both the Reynold's Number of flow and any friction factor (it effectively includes the latter). Theoretically it should also be the same for all sizes of a given component with similar geometry, but in practice this is not necessarily so because of variations in local geometry and so the true value can vary with different pipe sizes. Typical values are given in **Table IV**.

With laminar flow fitting resistance tends to be negligible since the theoretical loss in terms of equivalent length of straight pipe is equivalent to that of the actual length of the fitting regardless of its geometry (unless it presents a change in section producing a loss of velocity head, or introduces turbulence). In practice, however, it is general to use 'turbulent flow' values for fitting losses to be on the safe side for fluid viscosities up to about 100 centistokes. For viscosities above 100 000 centistokes fitting losses can generally be ignored, when flow is laminar. For intermediate viscosities, 'turbulent flow', fitting losses can be factored to arrive at a suitable value for turbulent flow – see **Fig 11.3.**

TABLE IV – RESISTANCE COEFFICIENT OF FITTINGS*

FITTING	PIPE DIAMETER – INCHES (MILLIMETRES)							
	¾ (20)	1 (25)	1¼ (35)	1½ (40)	2 (50)	3 (75)	4 (100)	5 (125)
Integral pipe bend (turbulent flow) bend 3 x D	Approximately 0.04 (all sizes)							
Integral pipe bend (turbulent flow) bend 4 x D	Approximately 0.025 (all sizes)							
Integral pipe bend (laminar flow) bend 3 x D	Approximately 0.1 (all sizes)							
Integral pipe bend (laminar flow) bend 5 x D	Approximately 0.06 (all sizes)							
Integral pipe bend (laminar flow) bend 10 x D	Approximately 0.04 (all sizes)							
Standard 90 deg elbow (screwed)	1.70	1.50	1.25	1.15	1.00	0.80	0.70	0.55
Standard 90 deg elbow (flanged)	–	0.45	–	0.40	0.38	0.34	0.32	0.30
Large radius 90 deg elbow (screwed)	0.90	0.75	0.60	0.50	0.40	0.30	0.25	0.20
Large radius 90 deg elbow (flanged)	–	0.40	–	0.34	0.30	0.25	0.21	0.18
Standard 45 deg elbow (screwed)	0.36	0.35	0.34	0.32	0.30	0.29	0.28	0.27
Standard 45 deg elbow (flanged)	0.34	0.33	0.32	0.31	0.29	0.28	0.27	0.26
Large radius 45 deg elbow (screwed)	–	0.24	0.23	0.22	0.21	0.20	0.19	0.18
Large radius 45 deg elbow (flanged)	–	0.22	0.21	0.21	0.20	0.18	0.18	0.17
Return bend 180 deg (screwed)	1.70	1.50	1.38	1.25	0.96	0.78	0.68	0.58
Return bend 180 deg (flanged)	–	0.43	0.40	0.37	0.34	0.32	0.30	0.28
Large radius return bend 180 deg (screwed)	–	0.80	0.70	0.60	0.50	0.40	0.35	0.30
Large radius return bend 180 deg (flanged)	–	0.42	0.39	0.36	0.30	0.25	0.22	0.20
Tee – line flow (screwed)	Approximately 0.9 (all sizes)							
Tee – line flow (flanged)	–	0.26	0.24	0.22	0.18	0.16	0.14	0.13
Tee – branch flow (screwed)	2.00	1.80	1.60	1.50	1.40	1.20	1.10	1.00
Tee – branch flow (flanged)	–	1.00	–	0.90	0.80	0.72	0.64	0.62
Screw-down valve (straight)	–	10.00	–	–	8.00	7.00	6.00	5.00
Screw-down valve (right angle flow)	–	5.00	–	–	4.00	3.50	3.00	2.50
Gate valve (typical) (screwed)	–	–	0.23	0.20	0.17	0.15	0.13	0.11
Gate valve (typical) (flanged)	–	–	–	–	0.30	0.23	0.15	0.13
Gate valve – ¼ closed	0.8 to 0.2 this range							
Gate valve – ½ closed	4.0 to 0.8 this range							
Gate valve – ¾ closed	16.0 to 2.0 this range							
Globe valve (screwed)	–	12.50	–	–	8.50	7.50	6.50	6.00
Globe valve (flanged)	–	12.50	–	–	8.50	7.50	6.50	6.00
Swing check valve (screwed)	–	3.00	–	–	2.00	2.00	2.00	2.00
Swing check valve (flanged)	Typically 2.0 (all sizes)							
Foot valve	Typically 0.8 (all sizes)							
Basket strainer	Typically 1.5 to 1.0 this range							

*Extracted from Handbook of Valves, Piping and Pipelines.

Flow Through Valves

Frictional losses through valves may be expressed in terms of equivalent length, a resistance coefficient (K); or a *flow coefficient* (C_V) in English units (but based on the US gallon) or the metric *flow factor* (K_V) with equivalents as follows:

$$K_V = 0.853\ C_V$$

$$C_V = 1.16\ K_V$$

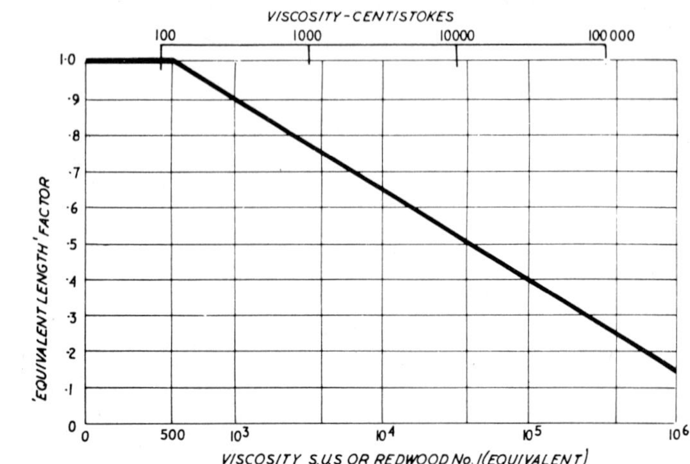

Fig 11.3

Working formulas then are:

$$\Delta P(\text{lb/in}^2) = \frac{\rho Q^2}{90 C_V{}^2} \text{ for Q in Imp gal/min}$$

$$= \frac{\rho Q^2}{130 C_V{}^2} \text{ for Q in US gal/min}$$

and ρ in lb/ft³

$$\Delta P(\text{bar}) = \frac{\rho Q^2}{223 C_V{}^2} \text{ for Q in litres/min and } \rho \text{ in tonnes/m}^3$$

Changes of Section

Changes of cross section of flow are involved at entries and exits, and also in expanding or contracting sections connecting two different diameter sizes of pipework. All such changes result in a loss of velocity head at the section and thus an energy loss which can be expressed in terms of a pressure drop or head loss.

Specifically, standard forms can be given a 'K' value or resistance coefficient, when head loss of pressure drop can be calculated from the formulas

$$\Delta P\ (\text{lb/in}^2) = 0.00673\ KV^2 \times \text{SG of fluid}$$
$$\Delta H\ (\text{feet}) = 0.01553\ KV^2$$

for V in ft/sec

Typical data are summarised in **Tables VA** and **VB**. For any additional information required on this subject, standard works on applied hydraulics shoul should be consulted.

TABLE VA – RESISTANCE COEFFICIENTS FOR CHANGES OF CROSS SECTION

	Ratio of Smaller Pipe Diameter to Larger Pipe Diameter								
	0.1	0.2	0.3	0.4	0.5	0.6	0.7	0.8	0.9
10° Taper				0.35	0.25	0.20			
20° Taper				0.15	0.12	0.10			
Sudden expansion	2.0								0.15
Sudden contraction	0.5	0.45	0.45	0.45	0.45	0.4	0.3	0.2	0.15
Inlet: Abrupt	Typically 0.5								
Gradual	Not less than 0.5								
Projecting tube	Typically 1.0								
Outlet: Abrupt	Typically 1.0								
Gradual	Not less than 0.12								

TABLE VB – CHANGES OF CROSS SECTION: APPROX. EQUIVALENT STRAIGHT PIPE LENGTHS

Pipe Dia.	1	2	3	4	5	6	8	10	12	14	16	18	20	24
Expansion d/D														
= 0.25	3	5	8	11	14	16	20	26	31	36	42	48	55	65
0.5	2	4	6	7	9	11	15	18	22	25	29	34	46	52
0.75	0.5	1	2	2	3	3	4	6	7	8	10	12	15	20
Contraction														
d/D = 0.5	1	2	3	3	4	6	8	10	12	14	16	18	20	26

Calculation of Total Pipework Losses

To summarise, evaluation of total losses involves the following procedure.

1. Divide the system into:-
   (a) straight lengths of pipe of equal diameter
   (b) fittings, each to be considered separately
   (c) sources of loss of velocity head, *eg* entry and exit and changes of section.

Then either:-

2. Find the equivalent straight lengths of fittings (1(b)), add to 1(a) and calculate effective pressure drop or head loss (*Note:* different pipe diameters must be treated separately).

Calculate other losses (1(c)) directly in the same units and add to 2, or –

(i) Calculate the pressure drop or head loss for the length of straight pipe(s). (*Note:* different pipe diameters must be treated separately).

(ii) Calculate fitting losses directly as individual pressure drops or head losses and sum. (Same units as (i)).

(iii) Calculate other losses (1(c)) directly. (Same units as (i).

(iv) Sum (i), (ii) and (iii).

Pipe Sizing

Calculation of system losses as described above is based on a specific pipe size, or implies that this is predetermined. An additional parameter is therefore necessary in order to arrive at what is likely to be a suitable pipe size, or at least provide a reliable first estimate. The most suitable parameter for this purpose is flow velocity, since an optimum flow velocity will compromise between excessive friction resulting from too small a pipe (flow velocity too high) and excessive cost resulting from the use of an unnecessarily large pipe (flow velocity too low).

An optimum flow velocity is generally derived from previous experience. Thus, in general, low velocities should be used on the suction side of the pump to keep frictional losses as low as possible and improve the NPSH. On the delivery side the fluid is under pressure and higher flow velocities can be used without trouble, so pipe sizing is far less critical.

Pipes are always assumed to have full flow when flow velocity can readily be determined from the flow rate (Q), *viz:*

$$V\ (\text{ft/sec}) = 0.15\ Q/D^2\text{, for Q in Imp gal/min}$$
$$= 0.133\ Q/D^2\text{, for Q in US gal/min}$$

where D is in inches

$$V\ (\text{metres/sec}) = 1.275\ Q/D^2 \text{ where D is in mm and Q in litres/min.}$$

It is generally more convenient to render this basic formula as a solution for pipe size consistent with a required flow rate and recommended value of flow velocity:

$$d\ (\text{inches}) = 0.4\sqrt{\frac{Q}{V}} \quad \text{for Q in Imp gal/min}$$

$$d\ (\text{inches}) = 0.36\sqrt{\frac{Q}{V}} \quad \text{for Q in US gal/min}$$

$$d \text{ (millimetres)} = 1.13 \sqrt{\frac{Q}{V}} \quad \text{for Q in litres/min}$$

TABLE VI – RECOMMENDED SUCTION FLOW VELOCITIES**

Pipe Bore		Water		Light Oils		Boiling Liquids		Viscous Liquids	
inches	mm	ft/sec	m/sec	ft/sec	m/sec	ft/sec	m/sec	ft/sec	m/sec
1	25	1.5	0.50	1.5	0.50	1.0	0.300	1.0	0.300
2	50	1.6	0.50	1.5	0.50	1.0	0.300	1.1	0.330
3	75	1.7	0.50	1.6	0.50	1.0	0.300	1.2	0.375
4	100	1.8	0.55	1.8	0.55	1.0	0.300	1.3	0.400
6	150	2.0	0.60	2.0	0.60	1.1	0.350	1.4	0.425
8	200	2.5	0.75	2.3	0.70	1.2	0.375	1.5	0.450
10	250	3.0	0.90	3.0	0.90	1.5	0.450	1.7	0.500
12	300	4.5	1.40	3.0	0.90	1.5	0.450	1.7	0.500
over 12*		5.0	1.50						

*General formula : pipe diameter (inches) = $\sqrt{\frac{\text{gal/min}}{10}}$

**Extracted from Handbook of Valves, Piping and Pipelines.

TABLE VII – RECOMMENDED DELIVERY FLOW VELOCITIES**

Pipe Bore		Water		Light Oils		Boiling Liquids		Viscous Liquids	
inches	mm	ft/sec	m/sec	ft/sec	m/sec	ft/sec	m/sec	ft/sec	m/sec
1	25	3.5	1.00	3.5	1.00	3.5	1.00	3.5	1.00
2	50	3.6	1.10	3.6	1.10	3.6	1.10	3.6	1.10
3	75	3.8	1.15	3.8	1.15	3.8	1.15	3.7	1.10
4	100	4.0	1.25	4.0	1.25	4.0	1.25	3.8	1.15
6	150	4.7	1.50	4.7	1.50	4.7	1.50	3.9	1.20
8	200	5.5	1.75	5.5	1.75	5.5	1.75	4.0	1.20
10	250	6.5	2.00	6.5	2.00	6.5	2.00	4.5	1.30
12	300	8.5	2.65	6.5	2.00	6.5	2.00	4.5	1.40
over 12*		10.0	3.00						

*General formula : pipe diameter (inches) = $\sqrt{\frac{\text{gal/min}}{20}}$

**Extracted from Handbook of Valves, Piping and Pipelines.

Suction Line Pipe Sizing

Recommended (maximum) suction flow velocities for various liquids are given in **Table VI** (see also **Fig 11.4**). A calculated pipe size will not necessarily be a standard pipe size (nor match the pump inlet flange size). The choice would then normally be the next standard size up, but if this is considerably larger

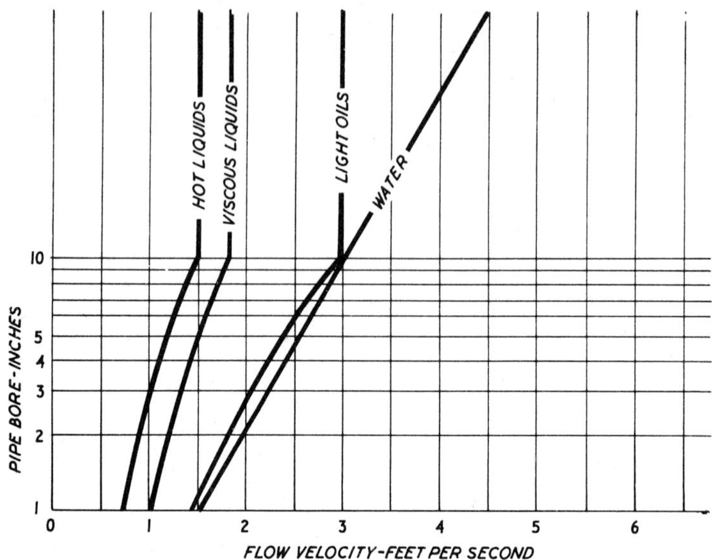

Fig 11.4: Recommended maximum flow velocities for suction pipes. Note these are arbitrary rather than specific recommendations. A simple rule commonly used for large bore pipes carrying water is:-

$$\text{pipe diameter required} = \frac{gal/min}{10} \text{ inches}$$

This is equivalent to a flow velocity of 5 ft/sec or 1.5 m/sec.

than the calculated diameter, or inconvenient to use, the nearest (smaller) standard pipe size can be checked. Calculate the flow velocity with this size and if this is not greater than 10% of the recommended (maximum) flow velocity, this pipe size should be suitable.

Delivery Pipe Sizing

Recommended (maximum) flow velocities for delivery flow for various liquids are given in **Table VII** (see also **Fig 11.5**). It will be appreciated that the higher flow velocities acceptable will allow the use of smaller pipe diameters on the delivery side and the selected value of flow velocity is largely arbitrary. Recommended values are based on what should provide an economic size of pipe (on first cost) whilst keeping frictional losses (and thus power losses) relatively low. The question of overall economics may, however, need analysing in more detail, especially where very long lengths of piping are involved, in order to arrive at a better estimate for optimum size.

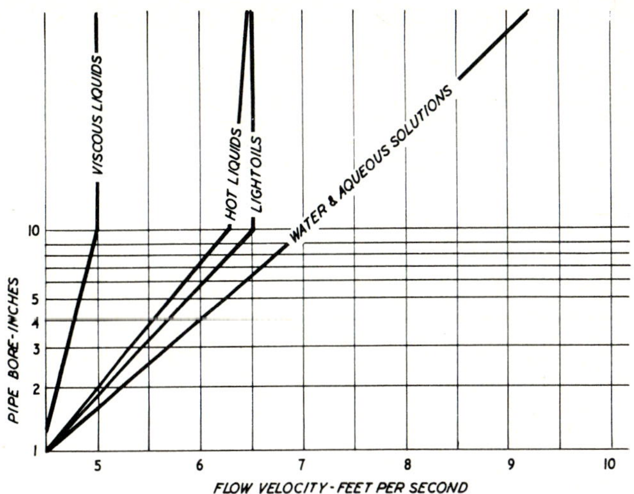

Fig 11.5: Recommended maximum flow velocities for delivery pipes. For large bore pipes carrying water the following method of preliminary sizing can be used:-

$$\text{pipe diameter required} = \sqrt{\frac{\text{gal/min}}{20}}\ \text{inches}$$

This is equivalent to a flow velocity of 10 ft/sec or 3 m/sec.

TABLE IXA – FLOW VELOCITIES FOR SLURRIES

Classification	Solids (Mesh Size)	Flow Velocity ft/sec	Flow Velocity metres/sec
Fines	Over 200	3–5	1–1.5
Sands	20–200	5–7	1.5–2.0
Coarse	4–20	7–11	2.0–3.3
Sludge	–	11–14	3.3–4.25

In other cases specific minimum values may be required for flow velocity on the delivery side – *eg* greater than the fall velocity when the fluid concerned contains solids in suspension.

Products which can prove critical in this respect include slurries and paper pulp (stuff) – see **Tables IX A, B** and **C.**

TABLE IXB – FLOW RATES (gal/min) FOR SLURRIES CONSISTENT WITH FLOW VELOCITIES OF TABLE IXA

Pipe Bore in	Fines	Sand	Coarse	Sludge
1	10	10–15	15–20	–
1½	15–20	30	40–50	60
2	30–40	50	60–80	90–100
2½	40–60	70–80	90–125	150–175
3	60–90	100–125	150–200	225–250
4	100–150	175–225	250–350	400
5	175–225	275–350	400–550	600–700
6	225–350	400–500	550–800	900–1000
8	400–600	700–900	1000	1500
10	700–1000	1000–1250	1500–2000	2500
12	100–1400	1500–2000	2500–3000	3500

TABLE IXC – FLOW VELOCITIES FOR STUFF & PAPER STOCK

Consistency	Optimum Flow Velocity	
Below 3%	10 ft/sec	(3 metres/sec)
3% – 8%	8 ft/sec	(2.5 metres/sec)

TABLE X – RECOMMENDED FLOW VELOCITIES BASED ON FLUIDS SG*

Pipe Diameter		Power Driven Pumps						Turbine Driven Pumps					
		SG = 1.0		SG = 0.75		SG = 0.5		SG = 1.0		SG = 0.75		SG = 0.5	
inch	mm	ft/sec	m/sec	ft/sec	m/sec	ft/sec	m/sec	ft/sec	m/sec	ft/sec	m/sec	ft/sec	m/sec
2	50	6.00	1.80	7.00	2.10	7.5	2.30	5.00	1.50	5.50	1.70	6.00	1.80
3	75	7.00	2.10	8.00	2.40	8.5	2.60	5.50	1.70	6.00	1.80	6.50	2.00
4	100	8.00	2.40	9.00	2.75	10.0	3.00	6.00	1.80	6.50	2.00	7.00	2.15
6	150	9.00	2.75	10.00	3.00	12.0	3.65	6.50	2.00	7.00	2.15	8.00	2.40
8	200	10.00	3.00	11.25	3.40	13.0	4.00	6.75	2.10	7.50	2.30	8.50	2.60
10	250	11.00	3.25	12.00	3.65	14.0	4.20	7.00	2.15	7.75	2.35	9.00	2.75
12	300	11.50	3.50	12.50	3.80	14.5	4.40	7.00	2.15	8.00	2.40	9.25	2.80
14	350	11.75	3.60	13.00	4.00	15.0	4.50	7.00	2.15	8.00	2.40	9.50	2.90
16 and over	400	12.00	3.65	13.00	4.00	15.0	4.60	7.00	2.15	8.00	2.40	9.50	2.90

*Extracted from Handbook of Valves, Piping and Pipelines.

Pipe Sizing by Specific Gravity

Since the power input required, and thus the energy cost of pumping, is directly proportional to the specific gravity of the product being handled, and an economic flow velocity varies inversely as pump speed, optimum sizes for delivery pipes (with centrifugal pumps in particular) are sometimes derived on this basis. Some recommended flow velocities related to product specific gravity and pump driver are given in **Table X.**

High Pressure Systems

With high pressure systems pipe sizing may be determined on the basis of a specified maximum acceptable *pressure drop* at the final delivery point. In high pressure fluid transport systems (as opposed to hydraulic circuits) relatively large pipe sizes are involved and recommendations are commonly based on *flow rates,* Some general data are:

Suction lines: maximum pressure drop 1 lb/in^2 per 100 ft run (0.25 bar per 100 metres), but dependent on available NPSH.

Delivery lines: acceptable pressure drops for different delivery rates are:

Flow rate		**Acceptable pressure drop**	
gal/min	**lit/min**	**lb/in^2/100 ft**	**bar/100 metres**
up to 100	up to 450	2 – 6	0.5 – 1.4
100 – 200	450 – 900	1.5 – 5	0.33 – 1.2
200 – 500	900 – 2250	1 – 4	0.25 – 1.0
over 500	over 2 250	0.5 – 2	0.1 – 0.5

Standard Pump Flanges

Pumps are normally fitted with flanges conforming to standard flange sizes, these branch sizes having been arrived at as optimum sizes consistent with the rated performance of the pump (rated delivery). The branch sizes are commonly taken as the parameter to decide pipe sizes and in a majority of cases this may prove a satisfactory solution. As a check, however, it is as well to calculate the resulting flow velocities from the approximate formulas

$$V\ (\text{ft/sec}) = \frac{\text{Delivery (gal/min)}}{6.9\ D^2} \quad \text{for D in inches}$$

$$V\ (\text{metres/sec}) = \frac{\text{Delivery (lit/min)}}{0.8\ D^2} \quad \text{for D in millimetres}$$

Alternatively **Table XI** may be used for rapid estimation of the order of flow velocities without resort to calculation.

If the actual flow velocities are found to differ markedly from typical or proven values for suction flow velocity, and delivery flow velocity, adjustment of the pipe sizes actually employed may be advantageous, *viz:*

(i) If the suction pipe velocity is too high a larger pipe size will improve NPSH and suction characteristics, or raise the suction lift attainable.

(ii) If the delivery pipe velocity is low, some saving in pipework costs may be realised by adopting a smaller standard size consistent with recommended values of V_d. This will, however, result in a reduction in working head.

(iii) If the delivery pipe velocity is high, an increase in pipe size will reduce the total head required, or increase the available head.

These effects will be most noticeable where the friction head accounts for an appreciable part of the total head. Where the total system head includes only a small proportion of friction head the size of pipe used on the delivery side is mainly significant only as affecting cost and cost of fittings.

12. Total System Head

DETERMINATION of total system head is a necessary feature for accurate selection of the size of the centrifugal pump required, since the capacity of such a type of pump is dependent on head. In the case of positive displacement pumps, determination of system head may establish the suitability of a particular pump type and also the actual capacity likely to be achieved, although such types are normally sized on a specific capacity rating alone (for a given speed). Positive displacement pumps are also less affected by change of operating conditions which may modify the total head.

Total system head may be defined completely by combining (*ie* summing) the static head and frictional and energy losses (expressed in terms of head), over a suitable range of capacities – **Fig 12.1**. The static head will be the same for all capacities, unless changing levels are involved, and so the static head curve will be a straight line at a constant head value. Friction head will be approximately proportional to the square of the capacity, this curve starting from zero with zero flow. The combination of the two curves, representing the *total system head,* will be a curve of the same form as the friction head curve, but displaced vertically by the amount of the static head.

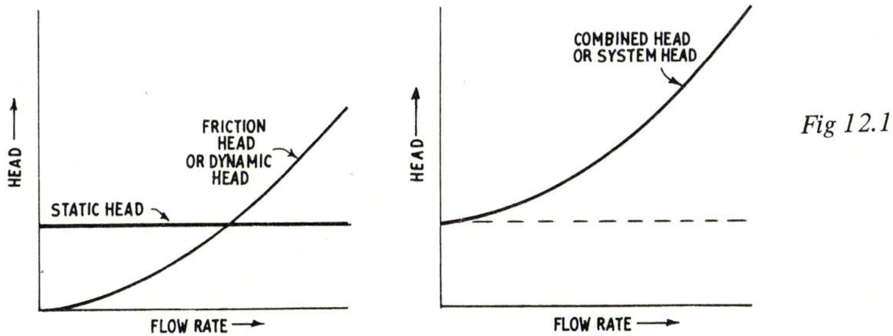

Fig 12.1

Such a complete plot of total system head enables the working point to be established of a centrifugal pump applied to the system analysed, simply by superimposing the H-Q curve for the pump on the total system head curve. The point where the two curves cross will represent the *working point* of the pump, or the H-Q value at which the pump will work – **Fig 12.2**. The same holds true

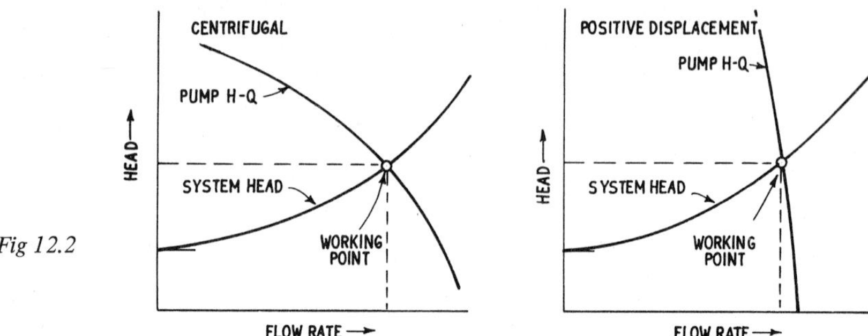

Fig 12.2

for any type of pump, the working point being established by the crossing point of the H-Q curve for that pump and the total system head curve, although the information is less significant in this case since it could be calculated directly.

In either case, providing the working conditions remain unchanged a single calculation for total system head at the capacity required will establish the working point, enabling a suitable size of pump to be selected. In the case of a centrifugal pump the optimum size would be a pump where the working point required corresponds to the working point for maximum efficiency of that pump, but, in practice, it is usually sufficient that the working point lies within the specified working envelope for the pump. In the case of positive displacement pumps it would normally be satisfactory to select the pump size to give the capacity required and then check, if necessary, that the total system head does not reduce the capacity below the required value. The pump size is not critical, for capacity could further be adjusted by the speed of the pump, although this obviously could not be applied to direct drives with constant speed motors.

Where working conditions are not constant, a plot of the total system head curve enables the effects produced to be investigated in detail. This plot need only extend over the range of changes likely to be experienced (*eg* on points established by a number of individual calculations), although analysis of complete curves will be used for the purpose of illustration.

Accurate determination of the total head is extremely important if the working point of a centrifugal pump is to be determined correctly, and from this the most suitable size of pump (which will ensure most efficient operation). An all too common error is to estimate total head on an approximate basis and then apply 'safety factors' to both head and capacity figures established. This is potentially dangerous since although a 'safety factor' applied to capacity or head separately may be quite realistic and yield limiting values still within the normal pump envelope or working range of a suitable size of pump, a combination of such factors may not. As a result of the apparent, but false, working point established, a different size of pump would be chosen which is not the optimum size for the particular service.

The usual error resulting in such cases is that the pump size chosen will be larger than necessary, adding to first cost and operating costs (power required). Also, although the pump has a larger capacity than necessary, its hydraulic performance may be unsatisfactory. Quite commonly, too, an oversize pump may be excessively noisy. On the other hand, a solution where a suitable size of pump is excessively noisy, is often to employ a larger capacity pump run at a lower speed to achieve the same working point. Thus noise alone is no criterion as to the suitability of a pump size; and equally the noisy 'oversize' pump would undoubtedly run more quietly and give more satisfactory hydraulic performance if its speed was reduced.

Regulated Discharge

The effect of partial closure of a throttle or regulating valve on the discharge side will be to increase and steepen the friction head curve, and thus the total system head curve. Each throttle position will thus establish its own working point, in the case of a centrifugal pump **(Fig 12.3a)**. A practical throttling range is also indicated by the limits of the working envelope of the pump, although it may be acceptable to employ further throttling if the lower efficiency is acceptable. Further information, which can be drawn from such a graph, is the head loss caused by throttling, represented by the vertical distance between corresponding points on the open throttle curve and the appropriate throttled curve.

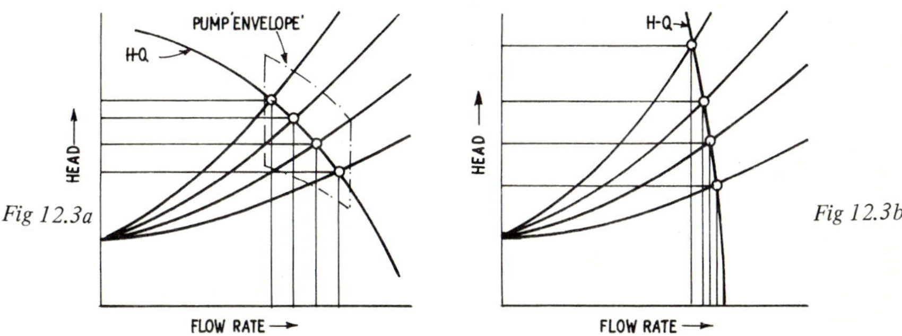

Fig 12.3a

Fig 12.3b

In the case of a positive displacement pump throttling is usually ineffective as a method of reducing delivery due to the near vertical form of the H-Q curve **Fig 12.3b**. Extreme closure may be necessary in order to produce a marked reduction in delivery which could lead to dangerously high pressure (high head) being developed by the pump. If throttling is adopted, therefore, protection for the pump should be provided by a relief valve opening at maximum safe pressure. This particular relief pressure will also establish the maximum possible capacity reduction.

Variations in Static Head

Quite commonly the static head may increase during a pumping operation, when the effect on the performance of a centrifugal pump is as shown in **Fig 12.4a.** An increase in static head displaces the working point of the pump to the left, with a corresponding reduction in capacity and an increase in head. Should the higher head operating point come outside the nominal working range of the pump a serious loss of efficiency could occur. It is thus desirable to select the pump so that the maximum increase in static head realised still leaves the working point within the pump envelope.

The effect on a positive displacement pump is far less marked and is generally negligible – **Fig 12.4b.**

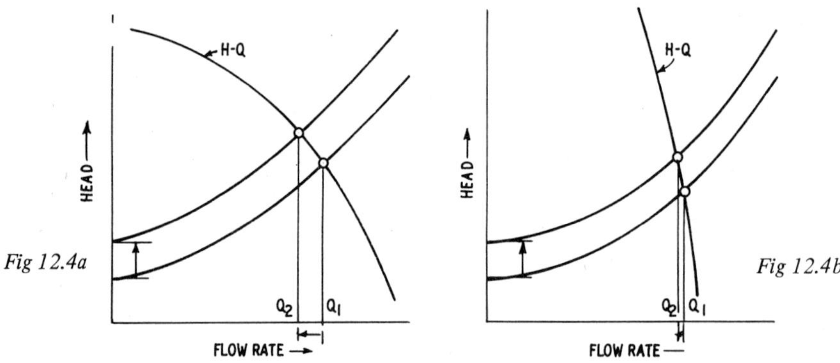

Fig 12.4a *Fig 12.4b*

Suitability of Centrifugal Pump Characteristics

The shape of the H-Q curve may determine the suitability of a particular centrifugal pump for a specific duty. **Fig 12.5** shows the performance potential of two pumps, one with a flat H-Q curve and the other with a steep H-Q curve, with both having the same working point for normal system head conditions. Any

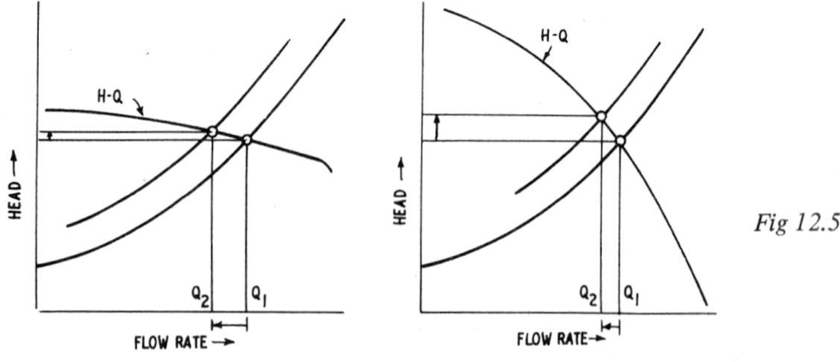

Fig 12.5

change in total system head will have a markedly different effect on the two pumps. Thus, an increase in total system head will result in a much greater reduction in capacity in the case of the pump with the flat H-Q curve; but equally will have relatively little change in head over a wide range of capacity. The pump with the steep H-Q curve, on the other hand, will show relatively little change in capacity over a considerable variation in total head. This may determine the type of pump best suited to a particular application.

The shape of the system head curve can be equally significant. Where the system head is mainly comprised of static head (*ie* short pipelines with little friction) the system head curve will be substantially flat. Any change in the static head may then produce markedly different results with centrifugal pumps having steep and flat H-Q curves, respectively – **Fig 12.6**. With a decrease in static head, in fact, the working point for the pump with the flat H-Q curve may approach the cut-off point. The type may, therefore, be ruled out for such applications.

Fig 12.6

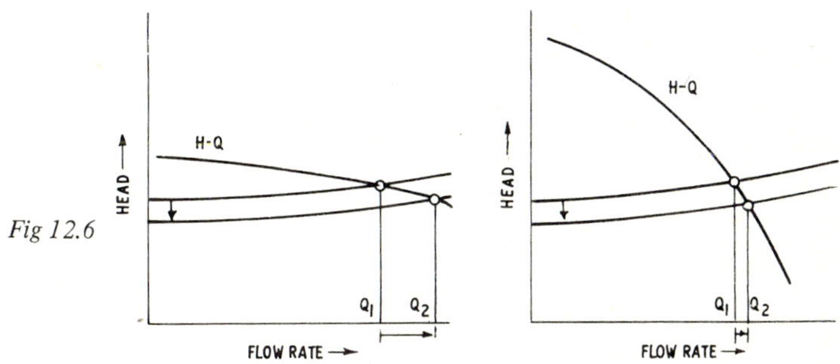

This, in fact, applies generally. If the system head curve is flat (mostly static head), pumps with a flat H-Q curve may be difficult to 'size' accurately, and can also give operating troubles with slight variations in total system head. For maximum stability under such conditions the steeper the H-Q curve of the pump the better, which favours rotodynamic pumps with high specific speeds for large capacities and positive displacement pumps for smaller capacities.

With a steep system head curve – **Fig 12.7** – the slope of the pump H-Q curve is generally less significant since the shift of the working point with change of head is much more restricted.

Particular attention must also be paid to matching system head characteristics and pump curves where the latter may exhibit unstable characteristics, or a sudden cut-off. The working point must be established on a suitable part of the H-Q curve where possible changes in the system head cannot shift the working point into an unstable or critical region of the H-Q curve if satisfactory hydraulic performance is to be maintained. Occasionally, too, the pump H-Q curve may have reflex characteristics with a dip and peak over a certain portion of the

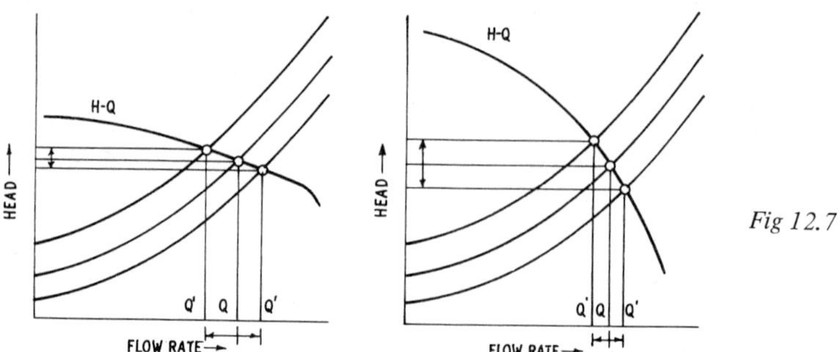

Fig 12.7

curve, *eg* this can sometimes happen with pumps of low specific speed which have H-Q characteristics bordering on the unstable when handling water and develop quite different characteristic shapes when handling more viscous fluids; also double-reflex curves are not uncommon with high specific speed pumps.

The above has discussed the relationship between system head curves (or calculations) and pump characteristic curves in general terms as a basis for selecting a suitable size of pump, (and also type of pump, where applicable). To extend this treatment further the most useful method of description is to consider typical individual systems.

No Static Head (Fig 12.8)

This is an application where the pump transfers liquid from one container to another, working with flooded suction and with no static lift involved. Losses are thus purely frictional (and energy losses at the inlet and outlet).

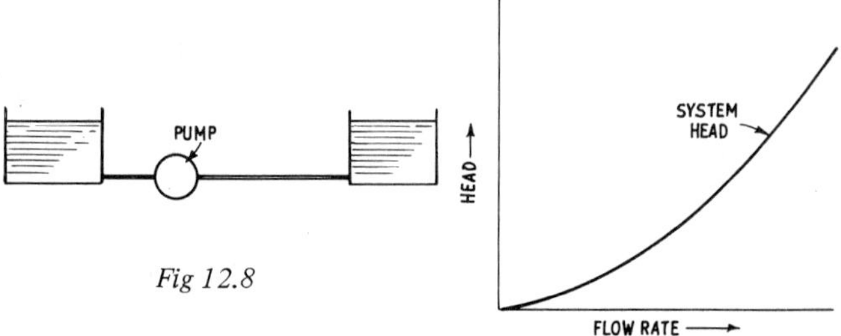

Fig 12.8

Normally a single calculation only would be required to determine the total head at the required capacity. This will establish the working point for 'sizing' a centrifugal pump. If regulation of discharge is to be incorporated, however, it may be necessary to calculate elements of the individual system head curves appropriate to different throttling conditions as in **Fig 12.3** in order to investigate whether the proposed throttling range lies within the working range of the pump, or on an acceptable part of the H-Q curve (*eg* not reaching an unstable point).

Calculation of friction head is based on standard procedure (see **Chapter 11**). Since the pump is normally located near to the supply, frictional losses on the suction side may be very small compared with frictional losses on the delivery side, (especially where the delivery line is long). In such cases it may be quite acceptable to ignore suction friction losses as these may amount to less than 1 or 2% of the total head. Where the delivery pipe is relatively short, total friction head may be quite small and the system head curve relatively shallow. This requirement favours the selection of a low head centrifugal pump with high specific speed (steep H-Q curve) for hydraulic stability; or a positive displacement pump, according to the actual capacity involved.

Mainly Static Head (Fig 12.9)

Here the duty involves a substantial static lift with a minimum of piping. The system head curve is substantially flat and the working point of the pump can be established by a single calculation. The effect of variations in total static head (*eg* due to falling level in the supply) can readily be estimated by simple arithmetic.

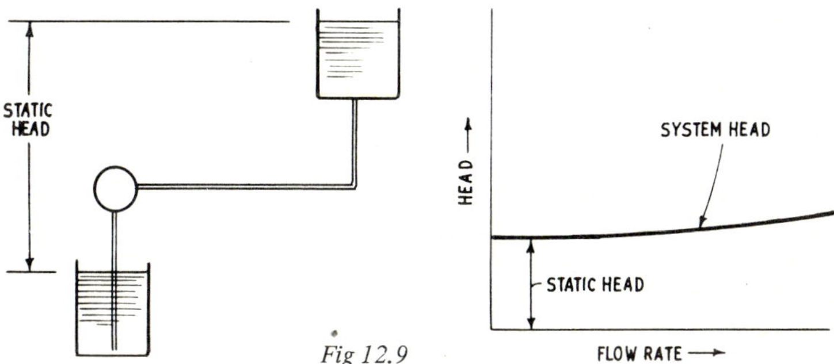

Fig 12.9

A common tendency is to assume that the friction head can be ignored in such cases but this is not necessarily true where (i) the static delivery head is only moderate; and (ii) pipeline size is small. To be on the safe side, therefore, the friction head should always be computed and added where a centrifugal pump is to be selected. The effect of a change of head should also be annotated to ensure that the working point always remains within the working range of the pump.

Combined Static and Friction Heads (Fig 12.10)

This is the most common application and is merely an extension of the previous example with the friction head accounting for a greater proportion of the total head. The working point is then established by a single calculation for the capacity required. The effect of varying conditions can then be investigated by further individual calculations, if necessary.

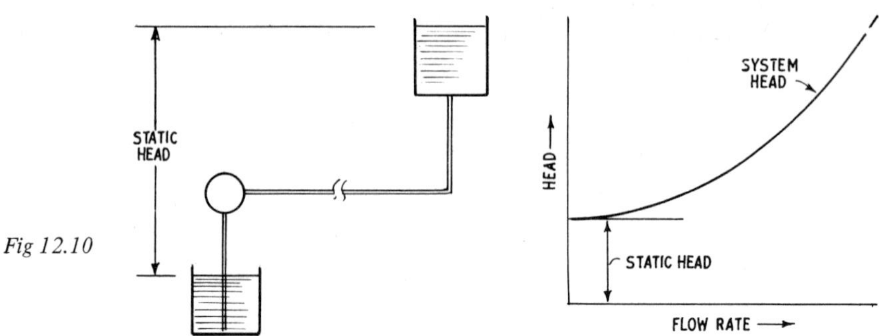

Fig 12.10

Gravity Head (Fig 12.11)

Fluid transfer is possible by gravity head alone as long as this head is greater than the head loss due to friction. System losses are thus frictional with a rising H-Q curve and a limit to discharge rate will be set by the point where the corresponding head loss equals the gravity head available. To produce a higher discharge a pump will have to be included to overcome the additional friction. The working point for the pump established by the frictional losses calculated for the higher rate of discharge will then be based on a total system head measured from H as zero. It should be noted that the capacity required will be the full discharge value as this amount will have to be passed by the pump.

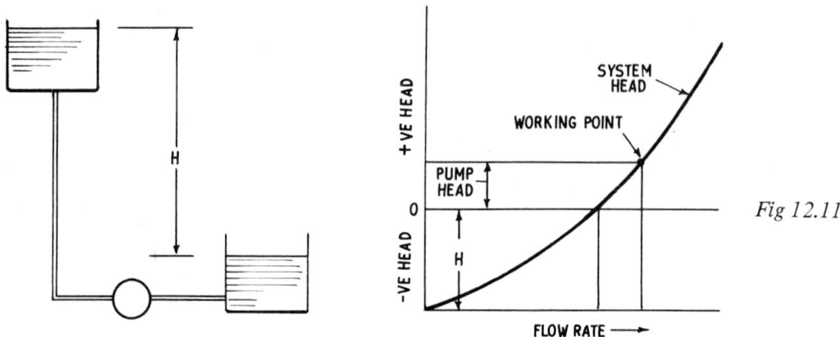

Fig 12.11

Parallel Delivery (Fig 12.12)

The method of arriving at the combined characteristics is to:

(i) Calculate the total system head for each pipe in turn.

(ii) Add together the discharges for the *same* total head. This gives the combined characteristics of the deliveries.

Fig 12.12a shows individual and combined characteristics plotted for two delivery pipes 1 and 2 having the same static head but different friction heads.

Fig 12.12b shows the individual and combined characteristics for two deliveries with different static heads. The working point is established on the combined characteristic curve, when the flow rates in the individual pipes can be determined by projecting across horizontally to cut the appropriate curves.

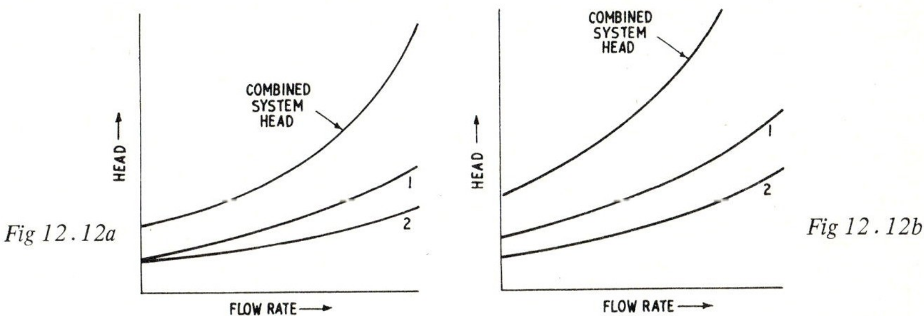

Fig 12.12a

Fig 12.12b

A similar method of analysis can be applied to diverted flow – **Fig 12.13.** Here part of the flow Q1 is diverted through Q2 as a separate delivery, with the balance of the flow Q3 feeding line 3. Head losses are first determined for lines 1 and 2 combined and plotted as a characteristic curve. Total head for line 3 is then plotted separately, starting at Q = Q2 since this quantity is already diverted before reaching line 3. The two curves are then summed to produce the combined system curve on which the working point is established, where Q1 = Q2 + Q3 is the capacity required. In practice, of course, the combined system head need only be calculated for this value of Q1 for constant operation of such a system.

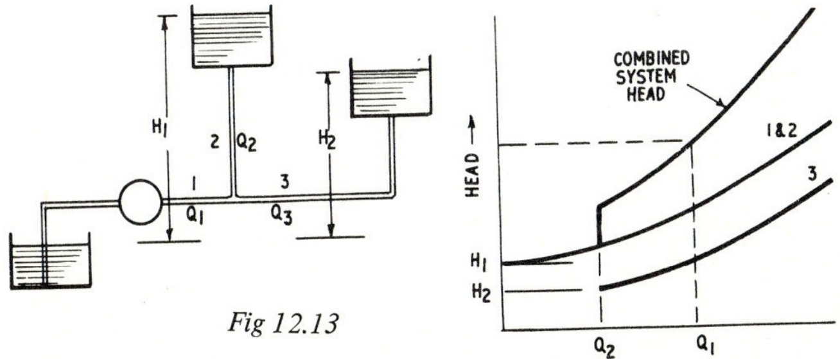

Fig 12.13

Paralleled Pumps (Fig 12.14)

When two or more pumps are operated in parallel the combined performance characteristics can be found by adding the capacities at the same head. Thus in the case of two identical pumps the combined characteristics can be found by doubling the capacity at various heads. The capacity at the operating point

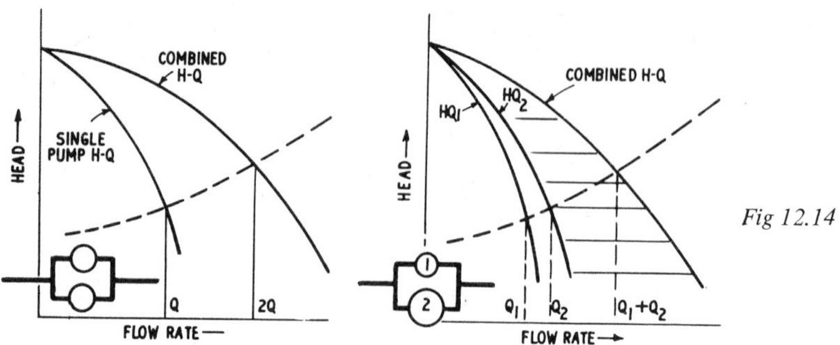

Fig 12.14

established by the intersection with the system head curve, however, will be less than twice the discharge obtained from a single pump working on its own in the system.

Where the individual pumps have different characteristics the combined characteristics are obtained by summation of the individual capacities at the same heads. The working points of the individual pumps are again established by the intersection of the system head curve.

The most common application of parallel operation is to provide an efficient method of regulating total discharge where demand is variable and centrifugal pumps are to be employed. Thus for small demand one pump only is run, with additional pumps cut in, as necessary, as demand increases. This is considerably more efficient for constant speed operation than regulating the discharge of one pump (or paralleled pumps) by discharge throttling.

It is a general characteristic when centrifugal pumps are operated in parallel that the flatter the H-Q curves of the pumps the more the discharge of the individual pump is reduced.

Paralleled operation may require investigation in a little more detail where the pumps are located far apart as each pump will then be working against a different system head, although discharging into a common pipeline. Another special case is where a centrifugal pump may be paralleled with a positive displacement pump. Here the characteristics of the positive displacement pump are represented by a substantially vertical line. The shape of the combined characteristics curve will thus be the same as that of the centrifugal pump curve but the combined capacity will be the sum of the two discharges. The effect is therefore for the combined H-Q curve to be shifted to the right by an amount equal to the capacity of the positive displacement pump.

Series Operation of Pumps (Fig 12.15)

Combined performance in this case is determined by adding the heads at the same capacity. Thus in the case of two identical centrifugal pumps the combined curve can be plotted by doubling the head ordinates for each value of

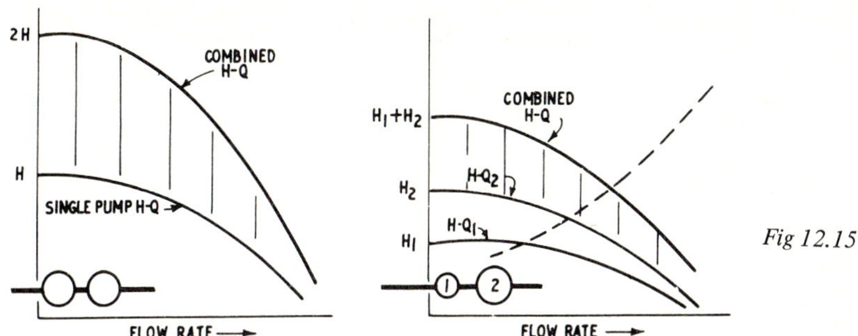

Fig 12.15

capacity considered. Where the characteristics are different the respective heads are summed at each capacity point to arrive at the combined characteristics. It follows that the flow rates must be the same, but the working heads on the individual pumps are only the same if the H-Q curves of the individual pumps are identical. The individual heads may be determined from the graphical analysis, *ie* by the working points of the individual pumps.

Series operation of centrifugal pumps has only the effect of boosting head. They would not normally be used for such duties as a multi-stage machine would normally be preferred where high heads are required. Series operation may be used for booster duties, however, where the performance of a main pump can be enhanced as necessary, by the use of a smaller booster pump in series circuit.

Adjustment of System Head

Whilst the total system head determines the working point of the pump it is possible to adjust system head, and thus the working point, by altering the parameters contributing to the total system head. Thus, replanning a run may alter the static head required, either by direct alteration of levels or the introduction of siphonic assistance.

Friction head is variable with flow rate, but is further variable with pipe size for a given flow rate. Thus the typical variation with capacity is based on a predetermined pipe size. For any given capacity friction head will be reduced if the pipe size is increased, and vice versa. The choice of an optimum pipe size is a separate problem,, and discussed in detail in **Chapter 11.** In exceptional circumstances, however, a further adjustment of pipe size may be justified in systems which have steep friction head characteristics in order to arrive at a more favourable working point.

Siphons

Fig 12.6 shows in simple diagrammatic form a pump system which includes a siphon head. In practice this may be occasioned by the necessity for the pipeline to negotiate an obstruction; or siphonic discharge may deliberately be introduced to reduce the total head on the pump during normal operation. Theoretically the maximum siphonic assistance possible is equivalent to 34 ft (10 m) of

Fig 12.16

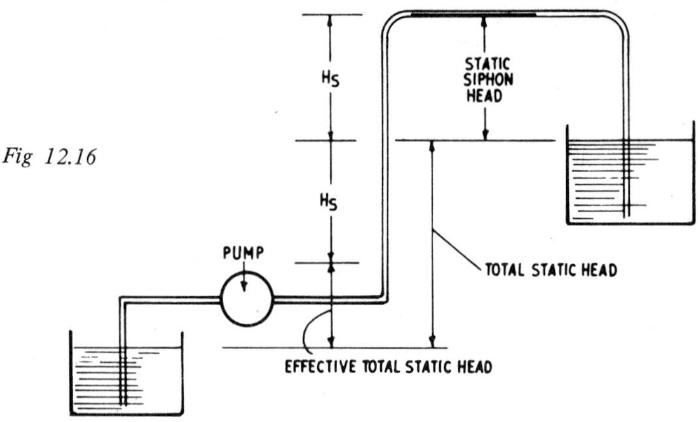

A typical application for a land drainage system is shown in **Fig 12.17**. The pump connects to a siphon discharge pipe with the invert being slightly above flood level and the outlet always being flooded. The actual pumping head is reduced in this case to the difference in levels between suction and discharge drains, plus frictional losses in the pipeline. A vacuum breaker is placed at the highest point in the pipeline, actuated by a paddle on which the water impinges to close the valve. The valve is biased towards the open position so that it will open as soon as water ceases to flow in the normal direction and thus break the siphon. This meets all the requirements for successful siphonic assistance, provided the flow is sufficient to fill the pipe cross section at the peak to ensure suitable priming conditions, and also that the pump is capable of developing a head equal to the height of the peak plus frictional losses in the line to the peak.

Flow velocity at the peak is somewhat critical and generally needs to be at least 5 ft/sec (1.5 m/sec). At the same time, liquid must not flow so rapidly from the discharge pipe that an excessively low pressure is developed at the peak. Pressure at the peak must always be greater than the vapour pressure of the

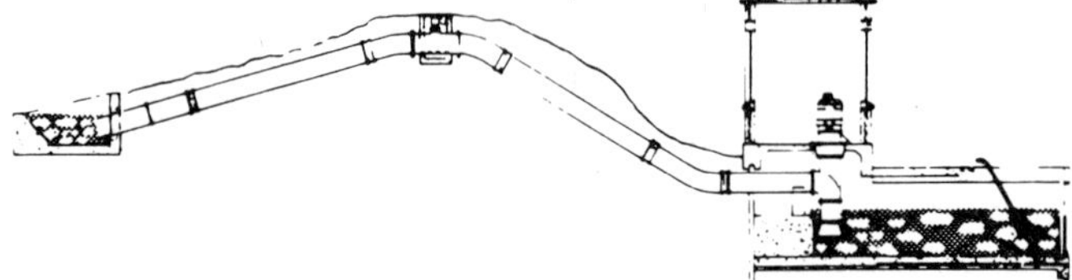

Fig 12.17: Axial-flow pump with siphonic discharge arrangement (Allen Gwynnes)

water, at sea level, and between 80 and 90% of this figure should be attainable in practice, with a correct design of siphon. Siphon losses will, however, vary with water temperature, flow velocity, barometric pressure and the amount of entrained air or gas present as well as the actual design of the siphon.

liquid, otherwise vaporisation may occur leading to the siphon becoming vapour bound. Equally, if vapour collects in the peak an air lock may develop, hence the desirability of incorporating an automatic air-relief valve at the highest point. Air entrainment can also modify the performance of the siphon, so as a safeguard the delivery pipe is always best terminated well below the fluid level.

A system employing two or more pumps discharging into a common main with siphonic assistance is shown in **Fig 12.18.** Each pump discharges through its own short individual siphon pipe into a surge tower and thence by a common main to the discharge point.

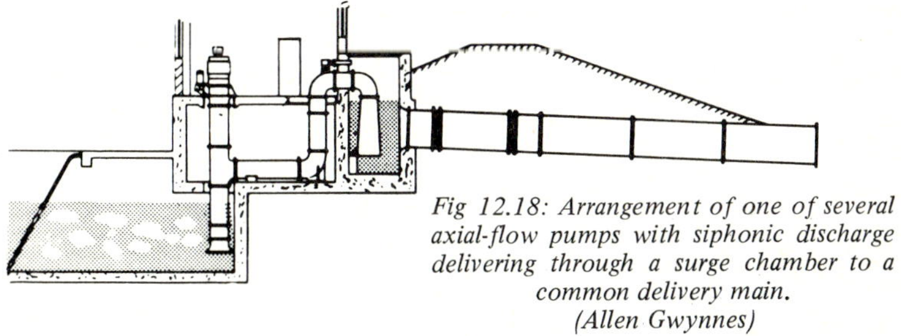

Fig 12.18: Arrangement of one of several axial-flow pumps with siphonic discharge delivering through a surge chamber to a common delivery main. (Allen Gwynnes)

Siphonic assistance may be employed to advantage for wet basin impounding and dry dock pumping, eliminating the sluice and reflex valves normally employed and reducing the size and cost of the pump required. Such a system is shown in **Fig 12.19** incorporating a hydraulically operated siphon breaking valve rather than the butterfly type valve previously mentioned.

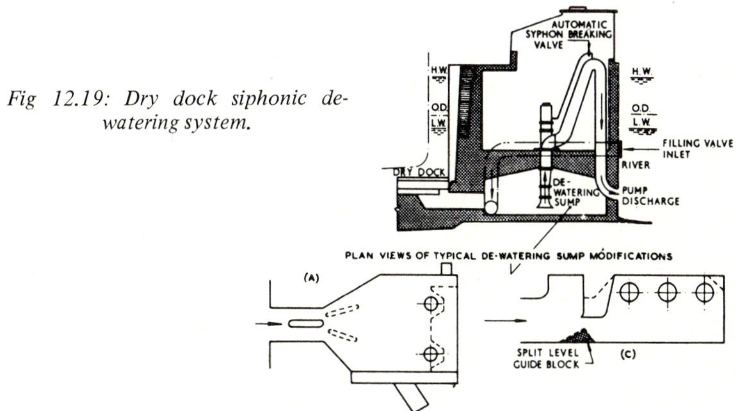

Fig 12.19: Dry dock siphonic de-watering system.

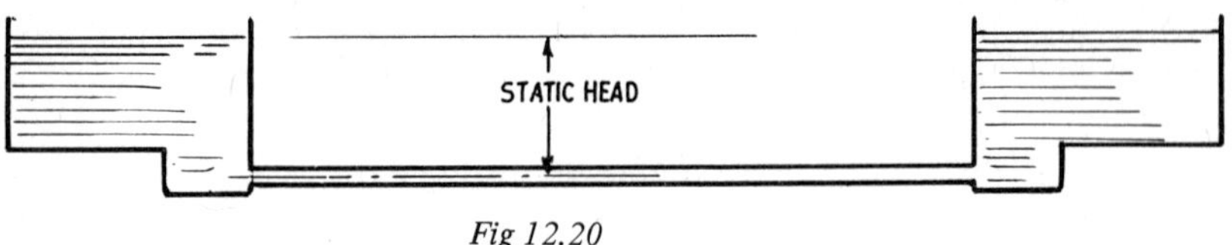

Fig 12.20

In some systems inverted siphons **(Fig 12.20)** may be required to negotiate the terrain involved, in which case similar siphonic assistance is available, the actual head loss being that of the frictional head of the connecting piping. In this case there is no need to provide air-relief. The main design factors involved are:

(i) Optimum pipe sizing to establish a minimum head loss where a long submerged run is involved.

(ii) Selection of pipes of suitable strength to withstand the full static head pressure developed where the run is deep.

13. System Design and Economics

THE OVERALL DESIGN of any pumping system, or system employing pumps for fluid transport, must be concerned with both hydraulic performance and economics. The larger the capacity of the system, and/or the more continuously it is worked, the more important economic factors become. Over a period considerably less than the life of the pump a difference in efficiency of only 1% may well exceed the initial cost of the pump and driver, which when applied to large pump sets can run into many thousands of pounds. The larger the scale of the undertaking, in fact, the more necessary it usually becomes to investigate alternatives, not only in pump types where choice is likely to be relatively restricted, but in pipework layout and detail design of the system.

Considering the problem generally, and starting with a fixed demand, the type of pump may or may not be more or less predetermined by the capacity requirements or the type of fluid to be handled. The system head, and particularly the friction head, may be regarded as somewhat variable at this stage. Any saving which can be realised in total system head represents a power saving, since it represents a reduction in hydraulic performance required, and thus power input required, *viz:*

$$\text{HP input required} = \frac{\text{gal/min x head x (ft) x SG of fluid}}{\text{3 300 x pump efficiency (\%)}}$$

Where pipelines are short and the head is mostly static, head pipe sizing is relatively unimportant, except for necessary considerations of flow velocity. The economics of the system are then more or less directly related to pump efficiency, and the efficiency of the driver as a fuel converter. The two together represent the overall efficiency of pump and driver. Data plotted in this way may well indicate the most favourable pump type/driver combination at the hydraulic power level output required, *eg* see **Fig 13.1**. Alternatively, individual combinations may be investigated and compared on the basis of overall efficiency at the specific capacity and head required, provided alternative drivers are feasible. In a majority of cases an electric motor is a more or less automatic selection for the driver, both as regards suitability and on account of lowest initial cost. The efficiencies achieved with electric motors is also substantially higher than that of prime movers.

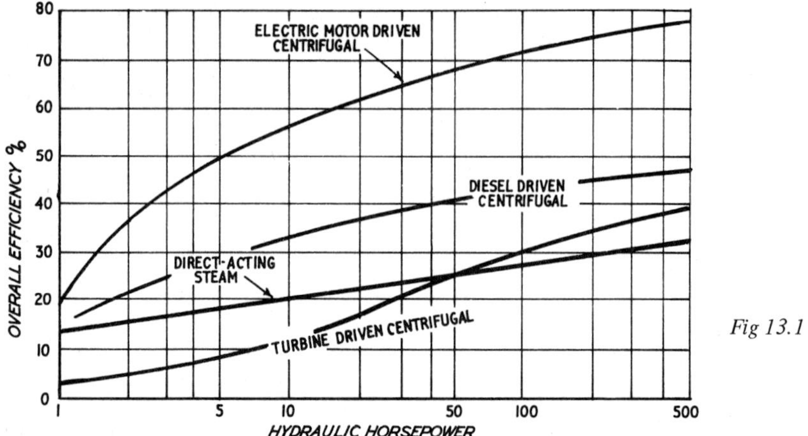

Fig 13.1

Some authorities advocate pipe sizing relative to the speed of the pump and specific gravity of the fluid. This is based on achieving economic flow velocities inversely proportional to pump speed and specific gravity. Such recommendations may differ from arbitrary pipe sizing on flow velocity considered with respect to friction head. Some typical data are summarised in **Fig 13.2**. It should be noted that these apply to delivery lines and to centrifugal pumps only. Further, any such data based on power costs are subject to variation with changing charges or fuel costs.

Methods of arriving at optimum pipe sizes as described in **Chapters 11** and **12** also do not necessarily result in optimum economic design for systems having a high friction head (*ie* long lengths of piping) since optimum sizing by flow rate does not take into account the cost of pipework. This question can be analysed separately in terms of:

(i) possible savings in static head
(ii) possible savings in friction head.

Possible savings in static head are intimately connected with a specific system projected and are concerned mainly with the routing of the pipework, which may be largely dictated by factors outside the system designer's control (although there are almost invariably alternatives to consider). A typical example is where an obstruction, such as a rise in the ground level, has to be negotiated. Carrying the pipeline over the obstruction will increase the static head of the system. Burying the pipe through the obstruction will increase the cost of installation. This increase in cost will, however, be offset by (i) a direct power input saving which can be calculated directly from the static head saved; and (ii) a smaller saving in actual pipe length reducing pipe cost and friction head,

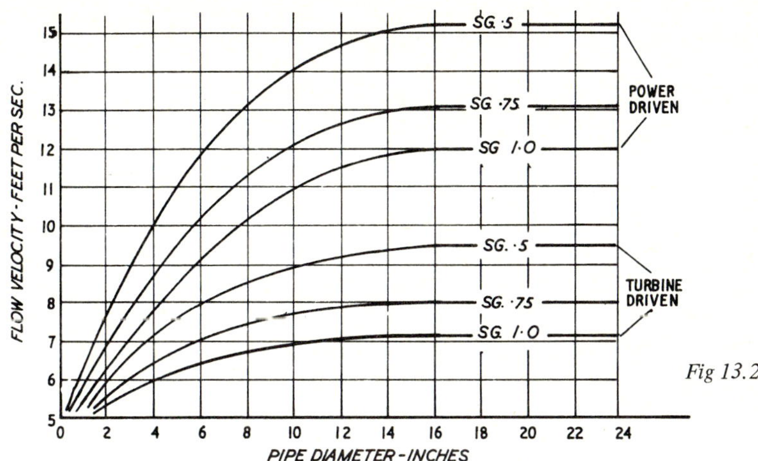

Fig 13.2

although these savings may be negligible by comparison with (i). The total degree of offset is calculable, although the possibility of other factors affecting the issue should not be overlooked.

In many cases involving bulk flow handling, questions such as the above may have a further factor, such as a definite limitation on the head available from the most suitable type of pump – *eg* an axial flow pump. An alternative scheme in such cases might be to split the duty between two (or more) pumps to provide 'lift' to a high tank or reservoir from one pump and bulk flow over substantially flat terrain from another pump; the use of a vertical multi-stage pump or an axial flow pump. Pipe sizing may also have to be adjusted to the hydraulic gradients involved, *eg* on downward sloping runs pipe friction needs to be high enough to ensure that the pipeline flow falls at all likely flow rates if a pump is also used in that line.

The larger the scale of the project, in fact, the more it becomes apparent that there is no 'exact' solution. Any solution must be a compromise to some extent and admits a number of possible alternatives which may involve the use of different types of pumps as well as different system layouts. Given the same problem, individual designers soultions would vary, based on their individual experience and judgement. Logically the best scheme would be the one giving the lowest *overall* cost per gallon of product pumped; although overall cost can be estimated, this is not necessarily a constant figure.

The *direct* cost of pumping can be rendered in terms of the cost of the power input, or directly as

$$\text{cost per gallon pumped} = \frac{\text{head x cost per HP x SG of fluid}}{\text{3 300 x pump efficiency (\%)}}$$

Additional costs are, however, inevitably involved, *viz:*

(i) Net depreciation of capital investment in
 (a) pump and driver
 (b) the system

(ii) Interest on any capital borrowed

(iii) Maintenance costs of
 (a) pump
 (b) driver
 (c) the system

(iv) Loss of performance through
 (a) wear on pump or driver
 (b) system corrosion

(v) Offset figures representing a recovery of higher initial costs which are not accounted for in the computation of direct costs. These may not be applicable since they may be fully accounted for in (iii) and (iv).

The *true* pumping cost is thus the direct pumping costs, plus these additional items suitably allocated over a period of working. Items (iii), (iv) and (v) can be regarded as fixed estimates. Items (i) and (ii) are not necessarily stable, hence the fact that the true pumping cost may change with time. This does not necessarily affect the validity of an initial analysis as charges or returns are equally capable of being varied. In practice, therefore, true pumping costs can be determined as a fixed figure, but one which may require periodic readjustment. This also takes care of any discrepancies in estimates covering the other additional charges.

Despite the fact that in certain services a pump may continue to work apparently indefinitely with suitable maintenance, all pumps should be costed on the basis of a specific working life. The life figure varies widely with the type of pump and service involved. In the case of a chemical or process pump working under extremely arduous conditions, pump life may be specified in months only. Typical life for other process and general duty pumps may be 1, 2, 5, 10, 15 or even more years.

An objection to continued use of a pump which is still (apparently) working satisfactorily, is that wear will inevitably result in loss of performance – *eg* it can shift the working point quite considerably in the case of a centrifugal pump and reduce efficiency accordingly – see **Fig 13.3**. Such a shift will tend to occur between maintenance periods when worn parts are replaced, but the longer the life of the pump the more difficult it will be recover full performance each time.

Economic analysis on this point can be based on power savings. At the point where power savings are greater than the 'write-off' cost of the pump in service, a replacement pump is obviously a better proposition – *eg* see **Fig 13.4**. Exactly the same method of analysis applies when considering the merits of replacing an existing pump with a more efficient one (or more efficient pump/driver combination).

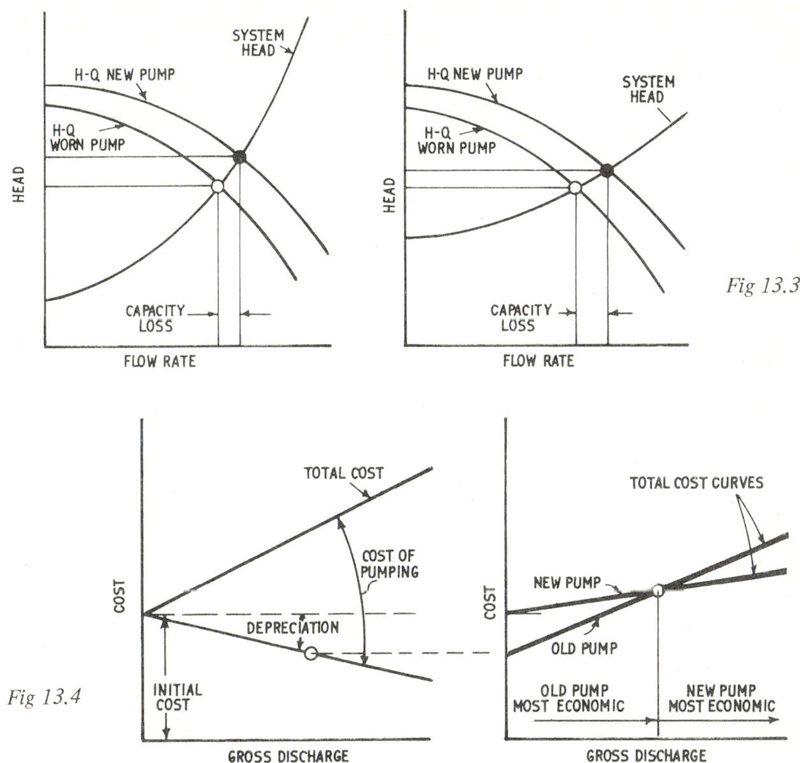

Fig 13.3

Fig 13.4

Optimum Pipeline Costing

Instead of friction head (and thus total system head) being determined for a single pipeline size (or a specific combination of sizes as indicated by optimum flow conditions), it can be instructive to calculate the friction head at the flow rate required for a variety of alternative pipe sizes – spot calculations only being necessary. The corresponding input powers required can then be plotted against pipe size; or more realistically the cost of the power plotted against pipe size – **Fig 13.5**.

A further set of data is then tabulated, relating pipe size to cost of pipework, *ie* material costs plus installation costs. Material costs will be approximately proportional to diameter, although actual figures are readily obtainable and should always be used. Material costs include those for pipes and fittings. Installation costs will probably not vary much over the diameter range investigated and, lacking specific figures, can be taken as constant. A curve showing total installation cost plotted against pipe size can then be plotted on the same graph.

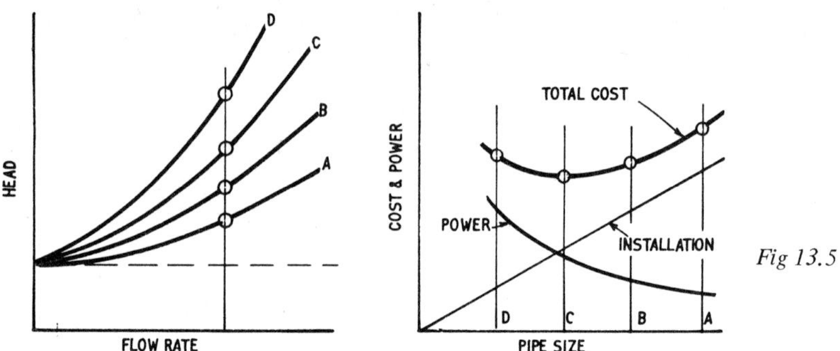

Fig 13.5

Pumping cost should be estimated as direct cost, based on a nominal or total gallonage selected so as to arrive at cost figures of the same order as total installation costs; or based on a specific period of working.

If the ordinates for the two curves are then summed and plotted as a single curve representing total cost (power *plus* installation) this curve will both indicate the minimum total cost which can be achieved and also the effect of differences in pipe size on total cost. Thus if the combined curve is substantially flat there will be little difference in choosing a range of pipe sizes, when it would be logical to employ the size originally arrived at from flow rate considerations. Equally, some adjustment of pipe size will probably have to be made anyway, from a theoretical diameter in order to utilise a standard pipe size. A non-standard size would inevitably increase the cost of the pipework and fittings, and thus the installation cost, raising the minimum cost point on the total cost curve. There may be exceptions to this rule, *eg* with long pipelines with high friction head, but these will clearly show on analysis – *eg* see **Fig 13.6.**

There may also be limitations imposed by acceptable friction head, which will automatically determine the minimum size of pipe which can be employed. Analysis is then restricted to analysis of the economics of larger pipe sizes only – **Fig 13.7.**

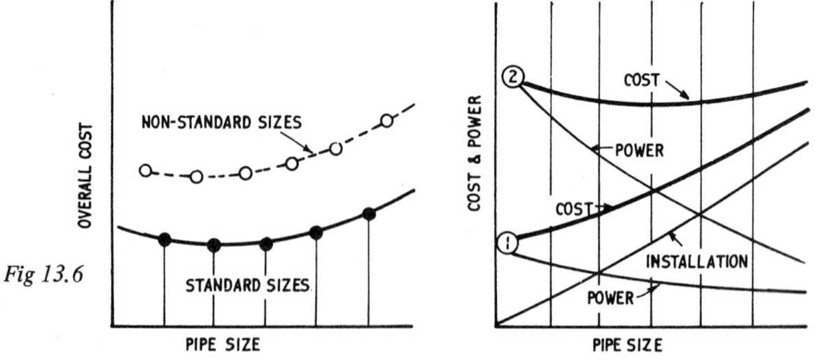

Fig 13.6

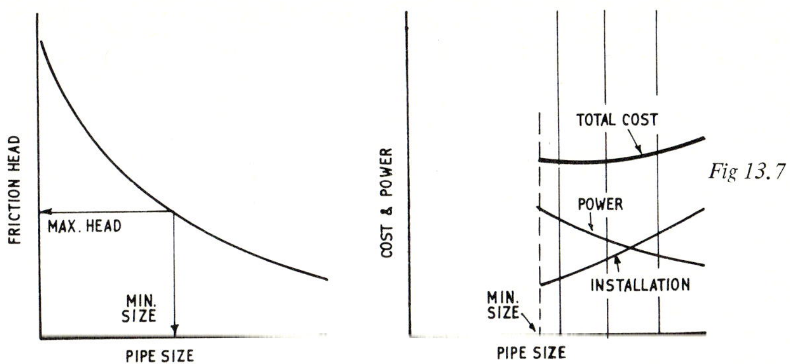

Fig 13.7

In any case, if the 'economic' pipe size determined differs appreciably from the initial size selected on the basis of flow rate, the effect on flow velocity should be investigated, particularly when the product being handled is contaminated with solids. An increase in flow velocity may lead to unwanted high local velocities at bends and erosion or erosion-corrosion problems; and a decrease in flow velocity may produce settlement and sedimentation problems. Certain parts of the system may also require minimum flow velocities to be maintained to ensure that the pipes maintain full flow – *eg* in syphons or negative inclines. In the main, however, pipe sizing is not usually a critical factor on the discharge side of a pump as far as flow velocities are concerned, so adjustment to an optimum economic size is usually possible. Suction line sizing should never be readjusted on this basis and flow conditions here are all-important; in any case the length of line is short and does not warrant 'economic' investigation. The same comment also applies to shorter lengths of delivery pipelines.

Variable Demand

Variable demand represents a shift in the working point of the pump which, apart from the mechanical control involved, can seriously affect the efficiency in the case of a centrifugal pump. This may favour the choice of a pump type which has variable delivery at more or less constant efficiency; or, if the capacity required is outside the normal range of such pumps, the use of two or more pumps in parallel, or a rotodynamic pump with more favourable characteristics.

Many systems for which a centrifugal pump is otherwise suitable may, however, have a variable demand, in which case a certain loss of efficiency may have to be accepted and part of the head or part of the capacity used for control purposes, using either discharge throttling or by-pass control. Both methods will inevitably result in power loss, so if economic regulation is of primary importance discharge regulation by speed control should be investigated first since this is less wasteful of power and loss of pump efficiency is usually markedly depleted.

Whereas speed control can be complicated, costly, or difficult to arrange using mechanical drives, the recent introduction of *frequency converters* covering a wide power range makes provision of speed control a simple process, and particularly advantageous when applied to centrifugal pumps. However a frequency converter for speed variation and control should not be assumed as an automatic answer. The optimum answer can only be obtained on the basis of which method of meeting the requirements of a variable demand system is the most cost effective. In many cases this may be simple throttling techniques.

The effect of discharge throttling on the working point of the pump is shown in **Fig 13.8.** In this case a head loss is accepted to provide capacity control, the cost being directly determinable by the head loss in the throttling valve. The actual loss need not be all that high, a head drop across the control valve of 30 to 40% of the frictional head loss at the working point generally providing good control. The steeper the H-Q curve the greater the power loss to achieve the same degree of throttling, and vice versa. If the full throttling range required can be achieved with the working point still remaining within the specified working envelope of the pump, then perfectly satisfactory working can be assumed. If not, then the pump should be sized so that the majority of its working time yields a working point within the pump envelope.

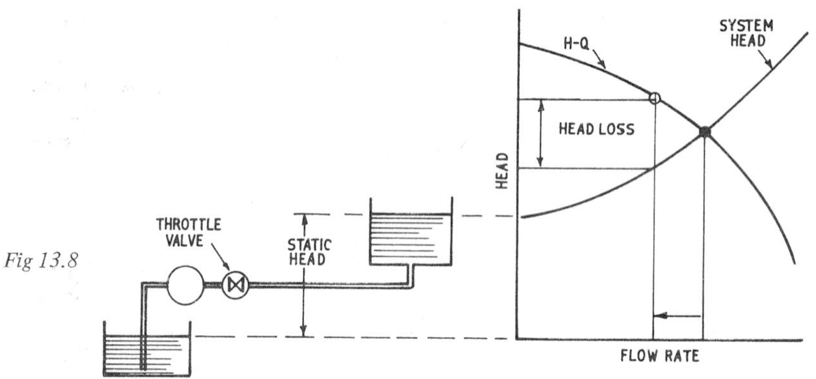

Fig 13.8

In the case of a pump with an unstable H-Q curve, discharge throttling can have a stabilising effect, provided the head loss across the control valve always *decreases* as the valve is opened – **Fig 13.9.** Under such conditions the valve will provide both stabilisation and effective capacity control. This effect is only likely to be realised, however, where the system head curve is quite steep, *ie* has a high friction head and relatively low static head.

The effect of by-pass regulation is shown in **Fig 13.10.** Here control is effected by using or 'by-passing' part of the capacity, or the pump in effect has two feeds. Rather than shifting the working point of the pump along the H-Q curve,

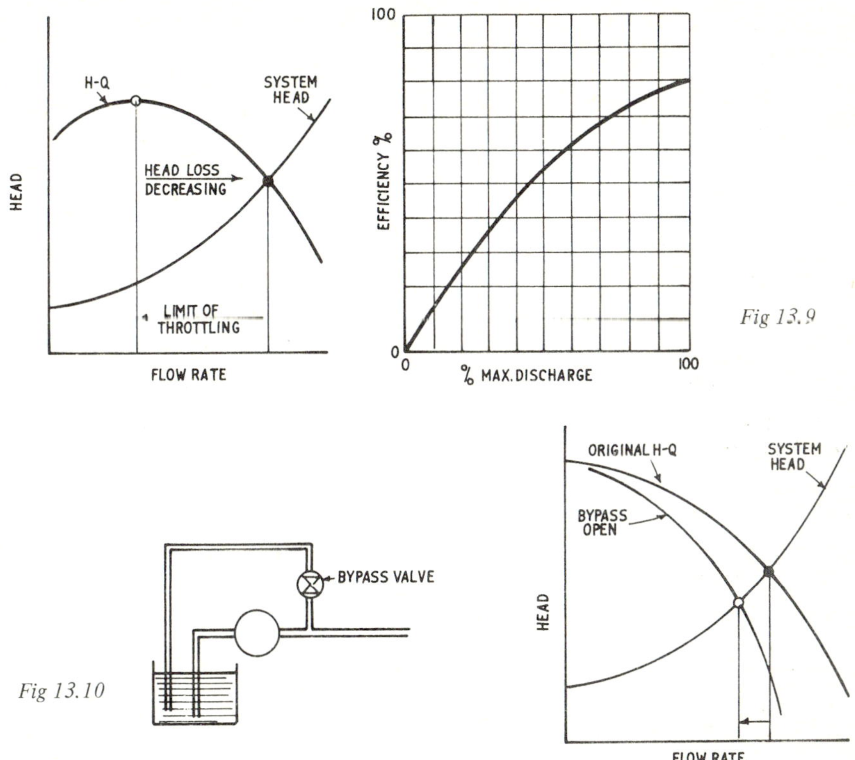

Fig 13.9

Fig 13.10

the effective H-Q curve is modified. as far as the main supply is concerned. By-pass regulation is mostly used on boiler feed pumps where the demand tends to vary over wide limits, and with higher specific speed rotodynamic pumps with steep H-Q curves.

Variable Demand and Economic Pipe Sizing

Each discharge rate will yield its own total system head figure for specific pipe sizes which, calculated as power costs and added to installation cost will yield a series of curves for total cost plotted against pipe size – **Fig 13.11**. Each delivery rate will then have its optimum pipe size, although the actual scatted of optimum pipe sizes may be relatively small. In general, the optimum solution would be to adopt the optimum pipe size indicated by the largest flow rate (nearest standard size); or the largest pipe size indicated, if this is different. Adjustment of pipe size to favour a particular throttled rate may, however, be better practice when the demand for this particular rate predominates as this will enable some of the cost of throttling to be recovered. This analysis is more

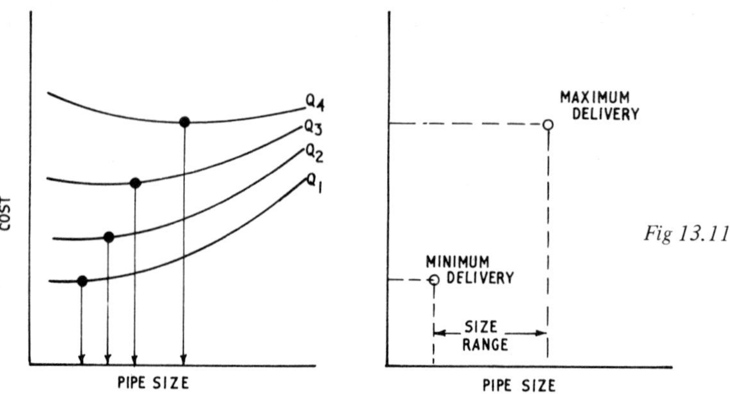

Fig 13.11

likely to be of academic rather than practical interest, however, since the main requirements with regulated flow are:

(i) to achieve the required variation in delivery to meet demand variations;

(ii) to realise good pumping efficiency at all demand levels.

Future Demand

With any extensive system demanding a heavy capital investment the possibility of changes in demand – particularly an increase in demand – need to be considered. This is likely to have a more significant effect on pipeline sizing than pump sizing and may well favour the adoption of apparently oversize piping. Part of the additional initial cost will then immediately be recoverable in power savings resulting from the lower frictional head.

From the pump standpoint, a small increase in demand could very likely be met by increasing the speed of the pump; and a large increase by the addition of another pump, with the system characteristics still remaining favourable. A solution which may appeal in certain cases where a substantial increase in demand is anticipated within a relatively short period is to adopt a larger pump initially, run at a lower speed, and subsequently increasing the speed and uprating the pump performance to meet a rising demand. This may provide a much cheaper overall solution than the purchase of a second pump.

Quite obviously, discussion of such matters in general terms has very distinct limitations. Individual system requirements present their own specific and individual problems and requirements. The main thing is to take all possible factors into account in order to arrive at the best possible compromise between hydraulic and economic requirements.

14. The Testing of Pumps

TESTS may be carried out on pumps for a variety of reasons, ranging from the determination of absolute performance of a particular design to spot tests in use to determine deterioration of performance through wear, *etc.* Major tests may demand the use of elaborate rigs, whilst for other purposes quite simple techniques may be satisfactory. The need for testing also varies with the type, and size of pump. Thus, fairly elaborate tests are needed to establish, or verify, the performance characteristics of rotodynamic pumps, and quoted performance must be achieved on production machines. With positive displacement pumps, characteristics are more fixed by the design and can be verified more simply. With small, low cost units, performance may only be nominal and justification may be unnecessary.

The majority of pump tests are concerned purely with the establishing of the hydrodynamic properties of a pump. More recently, however, this has been extended to thermodynamic determination of power input and internal efficiency, although the scope of this remains limited for normal practical requirements.

Tests may be considered under four separate headings, *viz:*

(i) *Laboratory tests* – these are mainly concerned with design development and experimental work where highly accurate measurement is required, and which also cover tests outside straightforward performance measurement, *eg* Research and Development work. This may involve:

- (a) Fundamental research into optimum hydraulic design using both aerodynamic and hydrodynamic test techniques.
- (b) Model tests.
- (c) Cavitation tests.
- (d) Investigation of noise, and noise reduction.
- (e) Materials research and technology under simulated operating conditions.

Laboratories may be part of a manufacturer's set-up or separate institutes. Larger laboratories with test rigs may also carry out acceptance tests.

(ii) *Works tests* – these are primarily concerned with production models and are used to verify design performance, verify differences produced by changes in detail design, and also provide spot checks on batch productions. Routine tests with limited measurement may also be carried out on individual models as part of the normal production programme.

(iii) *Acceptance tests* – these are usually applicable to larger machines (particularly rotodynamic pumps) where the contract of sale includes a guaranteed performance. Such tests are usually undertaken employing a rigid code specified by the National body concerned, which stipulated how measurements should be taken and the performance parameters computed. Acceptance tests may be carried out in the works, on the site after installation, or in a separate laboratory, according to circumstances.

(iiib) *Guarantee tests* – these are basically similar, but do not necessarily have to conform to any National Test Code, or meet a contracted performance other than that specified by the manufacturer as the pump's performance. It is desirable, however, that such tests should conform to accepted Codes in order to achieve consistency as regards guaranteed figures.

(iv) *Site tests* – involve tests on the complete installation or on individual pumps in the installation. Ideally these should be carried out at regular intervals to check the condition of the pump, *eg* for loss of performance occasioned by wear or corrosion. It is usually desirable in such cases that the test techniques be as simple as possible. Acceptance or guarantee tests may also be carried out on site after initial installation where it is inconvenient to do this beforehand.

Basic quantities to be measured are (i) discharge; (ii) head or pressure; (iii) rotational speed and (iv) power input. Suitable methods of measurement are summarised in **Table I**, but the appropriate Test Codes should be referred to for specific recommendations (when applicable). Particular attention must be paid to avoid errors in measurement which can readily arise from faulty location of pressure gauges and flow gauges, or equivalent devices.

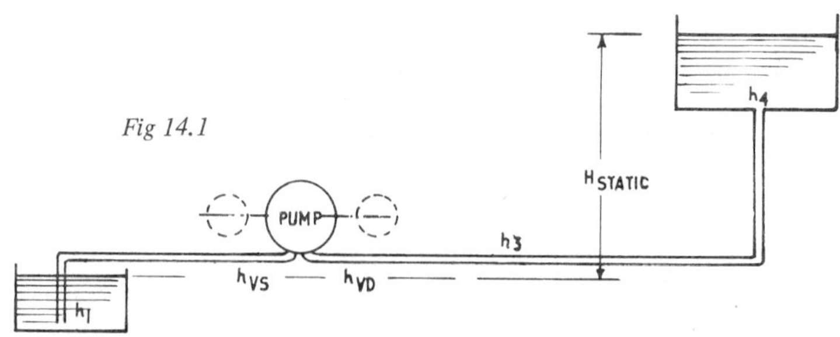

Fig 14.1

Basic Measurement of Performance

Consider the basic pumping system shown in **Fig 14.1.** The pump is called upon to raise liquid and then deliver it to a certain height. The difference between the

TABLE I – METHODS OF MEASUREMENT FOR PUMP TESTING

Quantity to be Measured	Suitable Instrument(s)	Remarks
PRESSURE	Manometers (mercury) (Not suitable for use with hot liquids)	Direct measurement of manometric head is given if one leg is connected to the pump suction branch and the other to the delivery branch.
	Differential manometers	Both spaces over the mercury columns should be filled with the liquid being pumped.
	Elbow meter (manometer)	Measurement of pressure drop at a bend.
	Spring-loaded pressure gauges (Bourdon tube, diaphragm and bellows types)	Former two widely used but need regular recalibration or calibration checks.
	Liquid column	Direct measurement of head; requires no calibration but is limited to heads of about 20–25 ft in practice.
	Piezometric devices Pitot venturi, nozzle or orifice)	Differential pressure measurement.
DISCHARGE	(i) Volumetric tank (ii) Weighing tank	Volume read directly or discharge quantity weighed and converted to volume.
	(iii) Gauging device (weir or orifice)	
	(iv) Flowmeters: inferential differential electro-magnetic	Flow rate derived from measured standing head. Piezometric devices
	(v) Special methods	May be adopted for discharge rates too high to be accommodated by standard means
SPEED	Revolution counter & stopwatch	Gives average speed only if speed is not constant.
	Tachometer	Gives continuous reading of speed.
	Optical transducer	Analogue or digital readout. Very accurate.
	Stroboscope	Gives instantaneous reading of speed with no physical connection to machine.
POWER INPUT	Torsion dynometer	Interposed between pump and driver.
	Dynamotor (swinging yoke motor)	Acts as pump driver.
	Inferential method	Deduced from electrical input to electric motor driver.
	Strain gauge (bridge)	Electronic measurement of instantaneous torque on shaft.

two liquid levels being defined directly as the total static head H_{static}. To perform satisfactorily, the total effective head generated by the pump must be greater than the static head by the amount of the losses involved. These losses are:

On the suction side: h_1 at the pipe inlet

h_2 representing the total frictional losses in the suction pipe.

On the delivery side: h_3 representing the total frictional losses in the delivery pipe.

h_4 representing the loss of velocity head at the outlet.

The total effective head to be generated is thus given by

$$H_e = H_{static} + h_1 + h_2 + h_3 + h_4$$

Under steady working conditions the total effective head can be determined by direct measurement of the pressures on the suction and delivery sides of the pump, *eg* with pressure gauges – see **Fig 14.2.** These gauges must be mounted on

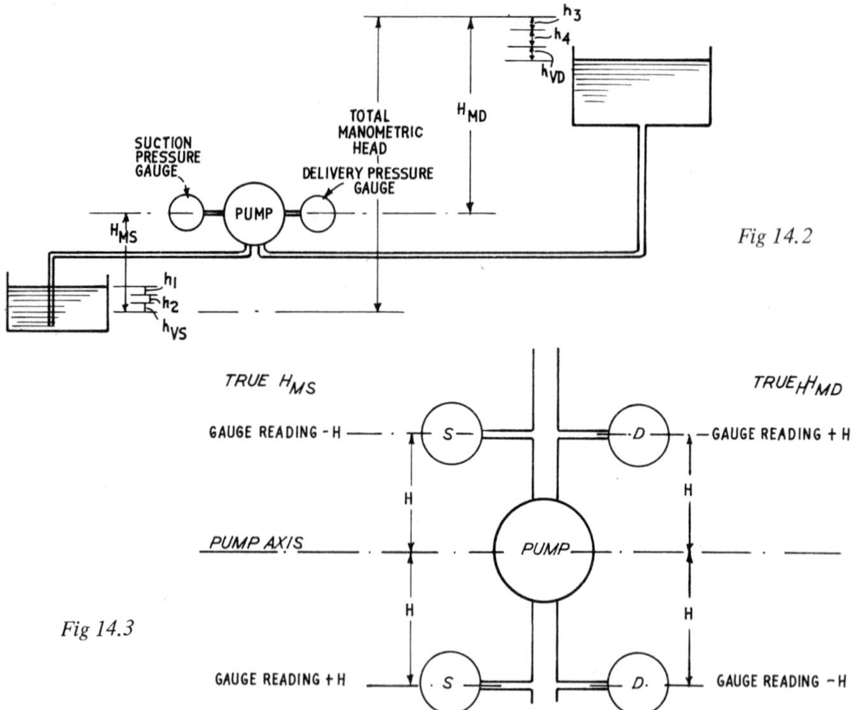

Fig 14.2

Fig 14.3

the same level as the centreline of the pump, otherwise they will themselves be subject to a head correction – see **Fig 14.3**. The fact that these gauges will read pressure rather than head is immaterial, but for the purpose of utilisation of the readings obtained, the pressure units must be converted into equivalent head of the liquid passing through the pump.

The gauge on the suction side of the pump will read effective suction pressure, which in terms of head equivalent is called the *manometric suction head* (H_{ms}) where –

$$H_{ms} = \text{geometric suction head } (H_s) + h_1 + h_2 + h_{vs}$$

The gauge on the delivery side of the pump will read effective delivery pressure, which in terms of head equivalent is called *manometric delivery head* (H_{md}) where –

$$H_{md} = \text{geometric delivery head} + h_3 + h_4$$

The *total manometric head* (also defined in **Fig 14.2**) is equal to $H_{ms} + H_{md}$.

Two other factors must now be considered – the velocity head at the suction flange of the pump (h_{vs}), and the velocity head at the delivery flange of the pump (h_{vd}).

The *total effective head* (H_e) is now given by

$$H_e = H_{ms} - h_{vs} + H_{md} + h_{vd}$$

The velocity heads (h_{vs} and h_{vd}) will be the same if the suction and delivery branches are the same size. If as is common the suction branch is larger than the delivery branch there will be a positive difference and the total effective head will be larger than the total manometric head. For most practical purposes, however, this difference can be ignored and the total manometric head (obtained from the pressure gauge readings) is taken as a measure of the total effective head. Such measurement of pump performance, of course, only holds true when the pump is running at a constant speed.

The same method of measurement can also be applied where the pump is working under flooded suction (*ie* with a positive static suction head). In this case the total manometric head will be given by –

$$H_m = H_{md} - H_{ms}$$

Simple Variable Head Tests

A simple method of providing variable head characteristics for works testing is to fit a valve in the delivery pipe. This valve can then be closed down to a degree necessary to produce the required resistance to yield a manometric head heading of the value required without the necessity of the pump having to discharge against a physical static head. This is particularly useful in the case of centrifugal pumps since the delivery head can be varied over a considerable range, if necessary, to establish a number of H-Q points with the pump running at constant speed. It is less suitable for rotodynamic pumps with steep H-Q characteristics, and generally unsuitable for propeller pumps.

In the case of positive displacement pumps, capacity is substantially independent of head, although in practice capacity will tend to decrease with increasing head because of increased internal slippage. Discharge throttling can be used to establish a true working head for simple flow measurement, although adjustment will be much more critical than in the case of centrifugal pumps. There is also the danger of building up excessive pressure on the discharge side of the pump if the valve is closed too much, although this could be safeguarded by the use of a relief valve in the system (if not already fitted to the pump as standard).

Throttling may also be employed on the suction side to regulate the suction head for test measurements, although here again the method has distinct limitations – Notably the possibility of the partially closed valve disrupting the normal suction flow characteristics. The risk of this can be minimised by locating the throttling valve as far away as possible from the suction inlet of the pump.

Simple Capacity Test

A simple set-up for a capacity test for small pumps is shown in **Fig 14.4**, suitable for a works or site test. In the latter case a tee-fitting provides a tapping point, normally blanked by closure of valve 1. For the capacity test both valves 1 and 2 are opened; valve 2 is then closed to shut off the delivery line and valve 1 adjusted to establish the required discharge head. Flow measurement is then taken from that point, the rise in level in the drum being measured over a given time and converted into capacity or delivery in that time.

An alternative method of establishing the measured discharge is by weighing. The drum is placed on a platform scale which is set to read a weight greater than that of the drum. The desired flow conditions are then set by adjustment of

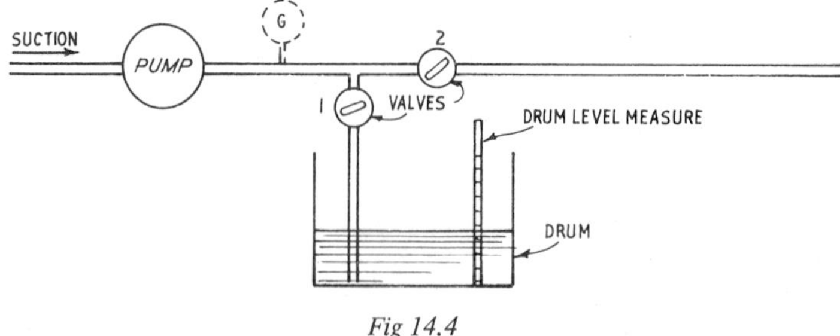

Fig 14.4

valve 1. As soon as the scale beam rises to the horizontal the stopwatch is started. Additional weight is then added to the beam and the stopwatch stopped as soon as the beam rises to the horizontal again. This gives the time for the discharge of a weight of liquid equivalent to the weight added to the beam, from which the volume can be determined, and thence the delivery of the pump. There are obviously alternatives to the actual method employed to arrive at the weight of liquid delivered in a measured time.

Open Circuit Test Rigs

Where the discharge is too large to be measured by the simple method described, a test rig will have to be employed. The discharge is then either determined by flow into a measuring tank, or by metering or measuring devices.

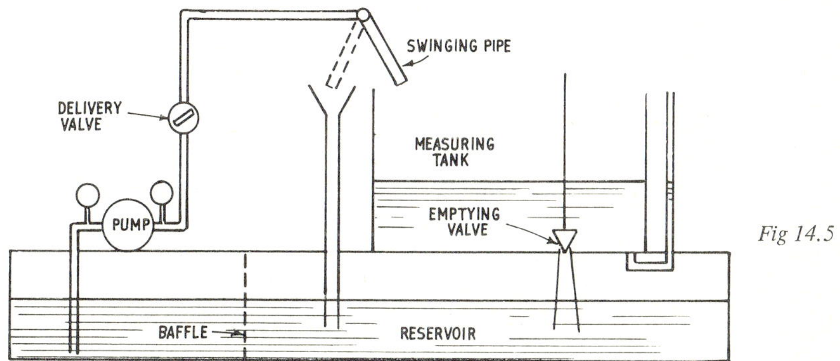

Fig 14.5

Fig 14.5 shows the simplest form of open-circuit rig employing a measuring tank and a suction reservoir, the latter being divided into two compartments with a baffle. A swivelling elbow enables flow to be bypassed directly back to the reservoir whilst the pump is being run up to speed, and the throttle valve adjusted to establish the required delivery head. Flow can now be directed into the empty measuring tank by swinging over the elbow, and back again to the bypass, after a given time. The delivery in this time is then read from a sight glass or measuring gauge fitted to the tank. Liquid is then returned to the reservoir by opening the tank outlet valve, this valve being designed to provide rapid emptying and a smooth discharge into the reservoir so that the next reading can be taken with the minimum of delay.

The main disadvantage of this form of test rig is that the liquid level in the reservoir falls during the test and thus the pump is effectively operating under varying suction head, and thus total head. The smaller the ratio between total head and fall in reservoir level the greater the effect on delivery. This may be overcome in various ways:

(i) Making the volume ratio of the reservoir and measuring tank large enough so that the fall in level of the reservoir is negligible.

(ii) Continual adjustment of the throttling valve so that the pump is always operating at a constant manometric head, *ie* by reduction of the delivery head in an identical manner to the increase in suction head developing.

(iii) The use of a double tank reservoir with an overflow partition and a feed pump to maintain a constant level in the suction chamber.

Two-tank systems can also be extended to the measuring tank with the advantage that they can be designed to provide almost continuous operation, one tank filling as the other is emptying.

Measuring tanks are generally the most suitable method of measurement for small pumps, although the rig is somewhat cumbersome. Alternative methods using weirs, orifices or flowmeters are generally less accurate for small rates of flow, and may also be limited in range (although this can usually be overcome by using multiple devices). They do, however, provide a relatively straightforward means of measuring large flow rates.

Open Circuit Testing: Weir Measurement

Weir characteristics are well established, or weirs can be calibrated by straightforward techniques. Basically a weir is a partial obstruction in an open channel over which liquid accelerates with a free liquid surface. Flow rate is simply related to the weir head by a coefficient, depending on the form of the weir – *eg* sharp edged or notched. Notched weirs are normally preferred – a triangular notch for greater accuracy with smaller rates of flow, and a rectangular notch for medium and large flow rates.

A typical rig employing a measuring weir is shown in diagrammatic form in **Fig 14.6**. Approach conditions are critical for accurate performance and so the pump outlet discharges into a stilling section separated from the main tank section by baffles. The measuring range with weirs is somewhat limited, unless provision is made to change from a triangular to a rectangular weir, or the two can be employed in parallel.

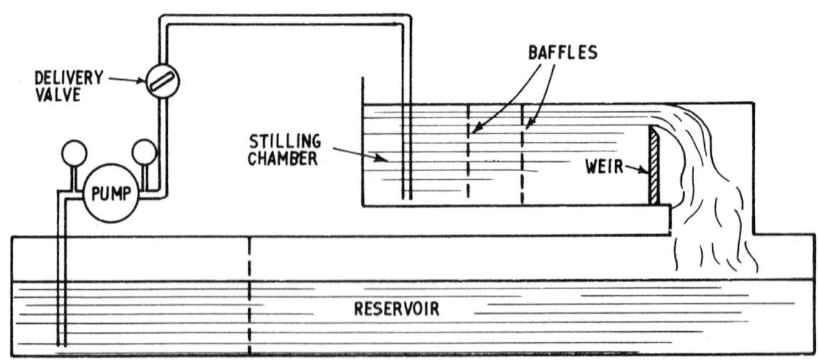

Fig 14.6

Open Circuit Testing: Orifice Measurement (Danaide Tank)

In the system shown in **Fig 14.7** the pump discharges into a tank fitted with a cylindrical baffle and with the base perforated with a number of gauging orifices. One or more orifices can be brought into use at a time, according to the flow rate concerned. The device can be calibrated to read discharge directly in terms of the head over the orifices and is thus similar in operating principle to the weir.

The danaide tank and the weir suffer from the common disadvantage that accurate reading of the head is difficult; also they may be subject to considerable time lag before the liquid level reaches equilibrium. Both, however, are widely used, particularly for laboratory testing.

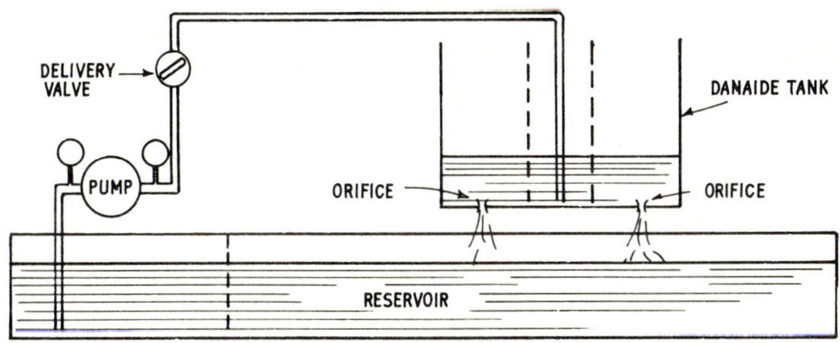

Fig 14.7

Open Circuit Testing: Venturi Meter

This method has the advantage of offering instantaneous readings, although the measuring range may be somewhat limited. The latter can be overcome by two meters in parallel circuits, with slightly overlapping ranges – see **Fig 14.8.** A practical disadvantage of venturi meters is that they are particularly sensitive to slight changes in flow conditions which could arise from a change in surface roughness in the upstream pipeline or in the meter passage itself, calling for the use of fully corrosion-resistant materials. Calibration checks should also be carried out at much closer intervals than those required with weirs or orifices.

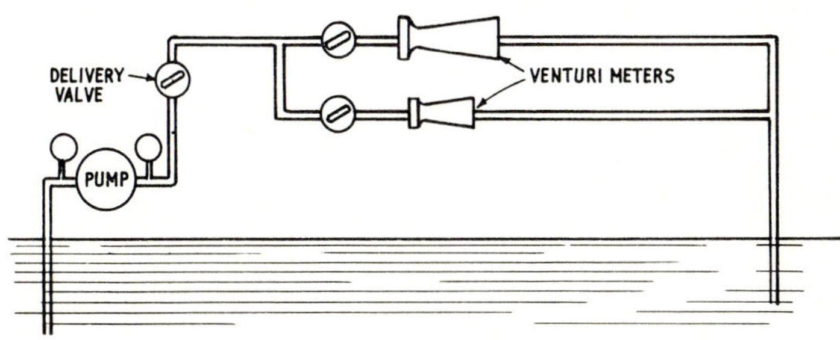

Fig 14.8

Closed Circuit Testing

Closed circuit test rigs are particularly applicable where high discharge rates are involved, since they avoid the need for large reservoirs and measuring tanks. They also provide a straightforward solution for testing pumps with special liquids, or at different temperatures, because of the constant volume of fluid involved. Basic disadvantages are the difficulty of regulating the rate of flow due to the rise in temperature of the working liquid which is inevitably bound to occur in the closed circuit.

A basic closed circuit rig is shown in **Fig 14.9.** Throttle valves are included to regulate both the suction and discharge whilst the circuit also includes a closed reservoir partially filled with air to allow for volumetric changes of the liquid. A flow measuring device is connected in a straight run of the delivery pipe and may be preceded by flow straighteners, if necessary.

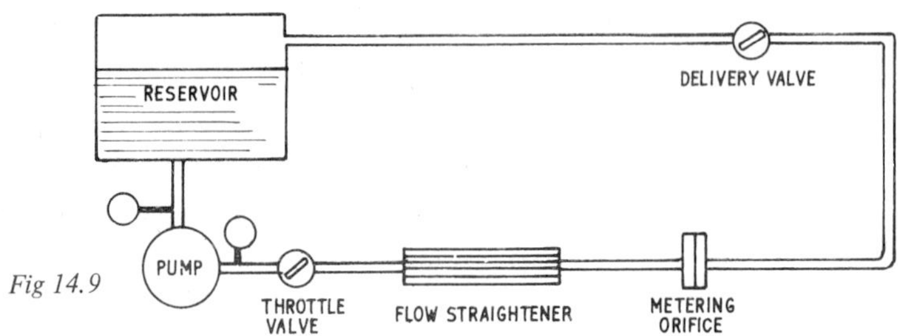

Fig 14.9

The basic circuit design may be varied in detail according to the type of pump being tested. Thus for testing pumps operating at high suction heads, throttling of the suction line is liable to give erroneous results and so this valve would be omitted (or left fully open). More realistic suction head conditions can then be given by tapping the top of the reservoir and connecting to a vacuum pump together with a means of regulating the suction conditions. Alternatively, the reservoir tapping can be connected to a compressed air supply to provide positive suction head for testing pumps normally operated with flooded suction. The same rig can provide both conditions (*ie* have both suction and pressure tappings for air).

For closed circuit testing with hot fluids (or where fluid heating is likely to cause discrepancies in measurement), cooling may have to be applied to the circuit (usually the reservoir), normally by the use of cooling coils. The cooling requirements may become quite elaborate with large rigs used for testing at high fluid temperatures.

Closed Circuit Site Tests

A bypass circuit may be installed close to a pump to facilitate periodic site testing. The bypass circuit then forms a loop as shown in **Fig 14.10** which is

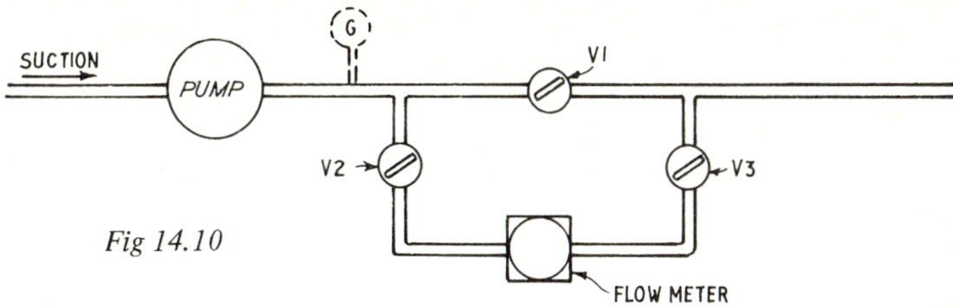

Fig 14.10

normally shut out of the flow line by closure of valves 2 and 3. Opening of these two valves and closure of valve 1 directs the full flow through the loop, when the flow rate can be indicated directly by the flowmeter in the loop.

This is not, of course, a true closed loop circuit but merely a bypass circuit employed to bring the flowmeter into the main flow to check pump performance. The reading is not a strictly true one since additional losses are involved in the loop and the manometric head cannot be adjusted to compensate via the effective discharge valve 3 which must be fully open. The additional losses are usually negligible, however, and in any case subsequent readings have only to be compared with an initial reading taken on the test loop to detect any deterioration in performance. A practical advantage for site testing is that the same meter can be used for testing a number of similar pumps in various circuits by connection to the loop valves 2 and 3 associated with each pump, *ie* a test loop is only completed by the connection of the meter.

The method can be extended to cover flow rates higher than those which can be handled by conventional flowmeters. In this case orifice plates are inserted in the discharge line and the entry side of the loop, the one in the discharge line having the greater number of holes – **Fig 14.11**. Pressure gauges are also fitted as shown. The throttle valve in the discharge line is then partially closed until the pressure gauged in the delivery and loop lines read the same. Flow rates though the delivery and loop lines will then be in proportion to the orifice areas in the respective orifice plates, *eg* the number of holes, if the orifice diameters are the same. Thus the total flow will be given by multiplying the indicated flow through the loop by the total number of holes in the two orifice plates.

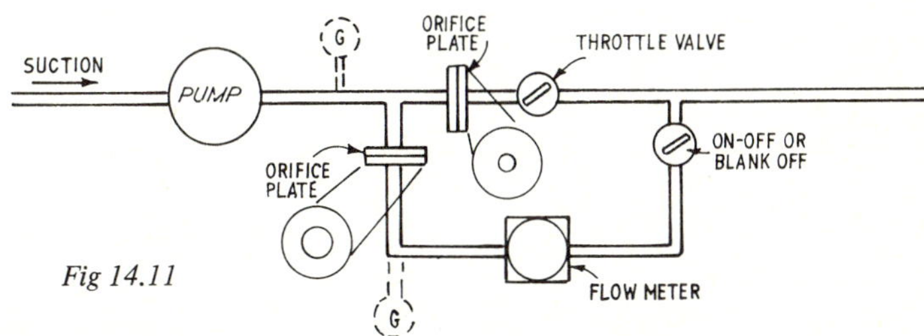

Fig 14.11

Although a practical system, the method does have certain disadvantages. Thus, to restore normal flow conditions the orifice plate needs to be removed from the discharge line, calling for disconnection of this line before and after testing. The loop circuit, of course, can be isolated with shut-off valves, or blanked off for normal system operation.

Cavitation Tests

Closed circuit rigs may also be used for cavitation tests, a basic set-up being shown in **Fig 14.12**. The reservoir is connected to a vacuum pump so that pressure can be progressively reduced during the test-run to the point where cavitation occurs. The onset of cavitation will be marked by an abrupt loss of total (manometric) head as read by the pressure gauges and a reduction in capacity (as read by the flowmeter).

Fig 14.12

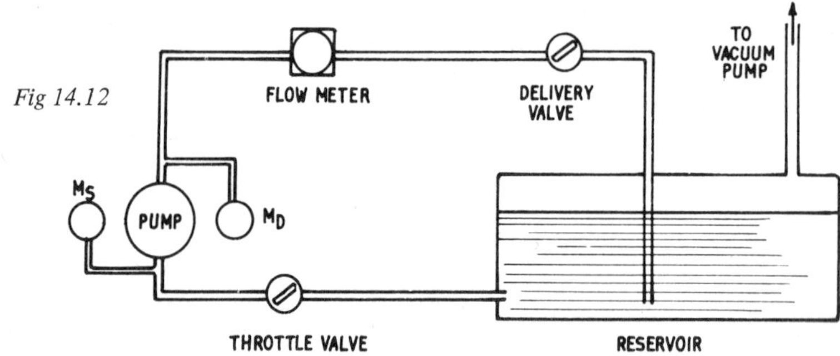

Test bed for pump cavitation investigation.

Air borne and structural noise measurement on a central heating circulating pump in the anechoic measuring room of the noise measurement centre. (KSB Pumps)

In practice a cavitation test rig may be rather more elaborate and include heat exchangers, both to control the temperature of the liquid, and enable cavitation tests to be run at different (constant) liquid temperatures.

Model Tests

Model tests using reduced scale models may be used to advantage to determine the hydraulic properties of very large rotodynamic pumps, and also the performance of complete installations or sections of such installations. Conditions of hydrodynamic similarity necessary for the accurate usage of model test data, however, are not always easy to obtain, or even to assess.

In general, for pumps of the same uniform shape, designating capacity by Q, head by H, and power input by P, and rev/min by N, and using the subscript m for the model –

$$Q = Q_m \frac{N}{N_m} \bar{d}^3$$

$$H = H_m \left(\frac{N}{N_m}\right)^2 \bar{d}^2$$

$$P = P_m \left(\frac{N}{N_m}\right)^3 \bar{d}^5$$

where

$\bar{d}$ = diameter ratio = full size impeller diameter/model impeller diameter

The relationship of the diameter ratio to head and capacity can also be extracted separately:

$$\bar{d} = \sqrt{\frac{Q}{Q_m}} \cdot \left(\frac{H_m}{H}\right)^{0.25}$$

or for speed ratio in terms of H and Q and equal specific speeds:

$$\frac{N}{N_m} = \sqrt{\frac{Q_m}{Q}} \left(\frac{H}{H_m}\right)^{0.75}$$

$$= \frac{1}{\bar{d}} \sqrt{\frac{H}{H_m}}$$

In practice, such extract translation may be appreciably modified by lack of true geometric similarity, *eg* differences in surface roughness, differences in relative thickness of impeller blades, *etc,* which can affect the actual hydraulic efficiencies which are realized. The greater the difference in size ratio the greater the likely difference in hydraulic efficiencies; or, in more complete terms, the greater the difference in Reynold's Number involved, the greater the likely discrepancy.

The Reynold's Number of a pump can be calculated in different ways, *viz:*

$$\bar{R}' = \frac{vD}{\nu}$$

$$\bar{R}'' = \frac{ND^2}{\nu}$$

$$\bar{R}''' = \frac{D\sqrt{H}}{\nu}$$

where

v = impeller velocity
D = impeller diameter
N = rev/min
ν = kinematic viscosity of fluid

Rather than the actual Reynold's Number it it is the ratio of the Reynold's Numbers which is important, calculated by the same formula. It is generally recommended that, for reasonable similarity $\bar{R}_m/R$ should lie within the limits 0.1–10, although this does not allow for geometric dissimilarities.

Various solutions have been advanced to show the likely relationship between size and relative hydraulic efficiency, which can be expressed in the general form

$$\frac{1-\eta h}{1-\eta hm} = \left(\frac{\bar{R}m}{\bar{R}}\right)^{a} \quad \left(\frac{1}{\bar{d}}\right)^{b}$$

or

$$\eta h = 1 - (1-\eta_{hm}) \left(\frac{\bar{R}m}{\bar{R}}\right)^{a} \quad \left(\frac{1}{\bar{d}}\right)^{b}$$

where

η_h = hydraulic efficiency, full size

η_{hm} = hydraulic efficiency, model

$\bar{d}$ = diameter ratio as before

Empirical tests indicate that the value of the exponential a is of the order of 0.1. The value of the exponential b would be zero for similar relative roughness, but in practice is usually some small positive quantity (typically of the order of 0.05).

Various semi-empirical formulas have been derived on these lines but none shows overall accuracy so that consistency of results is usually limited to a specific type of pump over a limited range. Some of these formulas are detailed in **Tables II** and **III.** Formulas for various other characteristics are summarised in **Table IV**.

Scale Effect on Maximum Suction Lift

The maximum suction lift attainable with a full size pump can be estimated from model data on the basis of

$$H_{1\,max} = H_b - H_{vp} - \delta H$$

and assuming that the same cavitation factor applies. This latter assumption is only justified –

(i) If the same liquid is used in each case, saturated with air to exactly the same degree.

(ii) The total heads (H) are the same for both the full size pump and the model.

Formula calculations are therefore justified for calculating the maximum suction lift attainable with a full size pump, where the cavitation factor is determined for the model pump under conditions conforming to (i) and (ii) above.

TABLE II – DYNAMIC SIMILARITY: BASIC RATIOS

Name	Force Ratio	Value ν
Reynold's Number	$\frac{\text{Inertia}}{\text{Viscous}}$	$\frac{\rho LV}{\mu}$
Froude Number	$\frac{\text{Inertia}}{\text{Gravity}}$	$\frac{V}{\sqrt{Lg}}$
Weber Number	$\frac{\text{Inertia}}{\text{Surface tension}}$	$\frac{V}{\sqrt{\sigma/\rho L}}$
Mach Number (gases)	$\frac{\text{Inertia}}{\text{Compressibility}}$	$\frac{V}{\sqrt{K/\rho}}$
Pressure Coefficient	$\frac{\text{Pressure}}{\text{Inertia}}$	$\frac{\Delta P}{\rho V^2/2}$

ρ = density
L = linear dimension
ΔP = pressure difference
K = compressibility
V = velocity
g = acceleration of gravity
μ = viscosity
σ = surface tension

TABLE III – DYNAMIC SIMILARITY: MODELLING RATIOS

Suffix m = model, f = full size

		Reynold's Number	Froude Number	Froude Number Distorted Model*
Force	$\frac{Fm}{Ff}$	$\frac{\rho f}{\rho m}\,\frac{\mu m}{\mu f}$	$\frac{\rho m}{\rho f}\cdot\left(\frac{Lm}{Lf}\right)^2$	$\frac{\rho m}{\rho f}\,\left(\frac{Lm}{Lf}\right)_H\left(\frac{Lm}{Lf}\right)^2_V$
Velocity	$\frac{Vm}{Vf}$	$\frac{Lf}{Lm}\cdot\frac{\rho f}{\rho m}\cdot\frac{\mu m}{\mu f}$	$\left(\frac{Lm}{Lf}\right)^{0.5}$	$\left(\frac{Lm}{Lf}\right)^{0.5}_V$
Angular velocity	$\frac{\omega m}{\omega f}$	$\left(\frac{Lf}{Lm}\right)^2\frac{\rho f}{\rho m}\,\frac{\mu m}{\mu f}$	$\left(\frac{Lf}{Lm}\right)^{0.5}$	
Flow rate	$\frac{Qm}{Qf}$	$\frac{Lm}{Lf}\cdot\frac{\rho f}{\rho m}\cdot\frac{\mu m}{\mu f}$	$\left(\frac{Lm}{Lf}\right)^{2.5}$	$\left(\frac{Lm}{Lf}\right)^{1.5}_V\,\frac{Lm}{Lf}$
Time	$\frac{tm}{tf}$	$\left(\frac{Lm}{Lf}\right)^2\frac{\rho m}{\rho f}\,\frac{\mu f}{\mu m}$	$\left(\frac{Lm}{Lf}\right)^{0.5}$	$\left(\frac{Lm}{Lf}\right)_H\left(\frac{Lf}{Lm}\right)^{0.5}$

*Vertical scale (V) larger than horizontal scale (H)

TABLE IV – CHARACTERISTIC QUANTITIES FOR PUMPS

Quantity	Value	Usual Symbol	Remarks
Width ratio	$\frac{\text{Impeller mouth dia.}}{\text{Impeller dia.}}$	γ	
Speed ratio	$\sqrt{2\,gH}$	ϕ	Also called spouting velocity
Flow ratio	$\frac{\text{Flow velocity}}{\text{Speed ratio}} = \frac{V}{\sqrt{2gH}}$	ϕ	
Head number (non-dimensional)	$\frac{gH}{D^2 \times (\text{rev/min})^2}$	hc or Hc	
Discharge number (non-dimensional)	$\frac{Q}{D^3 \times \text{rev/min}}$	qc or Qc	
Specific speed (true)	$Ks \cdot \frac{\sqrt{Q} \times \text{rev/min}}{H^{0.75}}$	Ns Ns	Value of Ks depends on units employed
Specpfic speed (nominal)	$\frac{\sqrt{Q} \times \text{rev/min}}{H^{0.75}}$	Ns	Value depends on units employed
Shape number*	$K \frac{\sqrt{Q} \times \text{rev/min}}{H^{0.75}}$	–	Value of K depends on units employed
Head coefficient	$\frac{gH}{(\text{peripheral velocity})^2}$	–	Non-dimensional
Flow coefficient	$\frac{V}{\text{peripheral velocity}}$	–	Non-dimensional

***Note:** This is basically the same as specific speed but employing a different factor. The shape number will therefore differ from the specific speed number, both calculated in the same units.

Testing with Air

It is sometimes more convenient to employ air for testing model pumps, instead of water or a liquid medium, particularly as this can substantially lower the cost of both the models and test rig and involve considerably lower energy levels (roughly 1/800th). The main disadvantages are the large differences in Reynold's Numbers which may be involved and the effect on efficiency; and the low pressure levels to be measured which can result in difficulty of reading. Both these disadvantages can largely be overcome by using compressed air with good compatibility once $\overline{R}_m/R$ is of the order of 1:15 (bearing in mind that these numbers are calculated for air and water, respectively).

P	*Pump*
M	*Motor*
B	*Measuring orifice plate*
K	*Cooler*
S	*Gate valve*
E	*Degassing tank*
D	*Dome*
o–.–o	*Measuring point*
V	*Globe valve*
VP	*Vacuum pump*
PP	*Compressed air pump*
o	*Gas bubbles*
------	*Strainers*

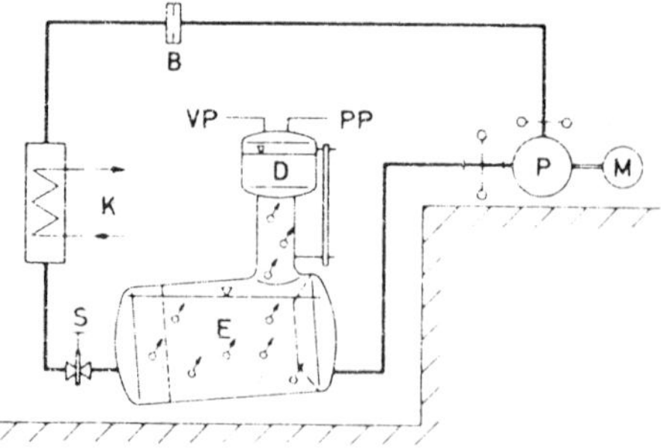

'Closed' test bed (diagrammatic). (KSB Pumps)

Example of a pump test house.

The basis of a simple rig for air testing using air at atmospheric pressure is shown in **Fig 14.13**. A simple manometer can measure the inlet and delivery pressures, and thus determine the manometric head; whilst a suitable flowmeter can determine the flow or quantity delivered. Power input can be measured by a dynamometer motor drive, although it is unlikely that any satisfactorily accurate measurement of power can be obtained at the very low energy levels involved. It is readily possible to measure both H and Q with sufficient accuracy

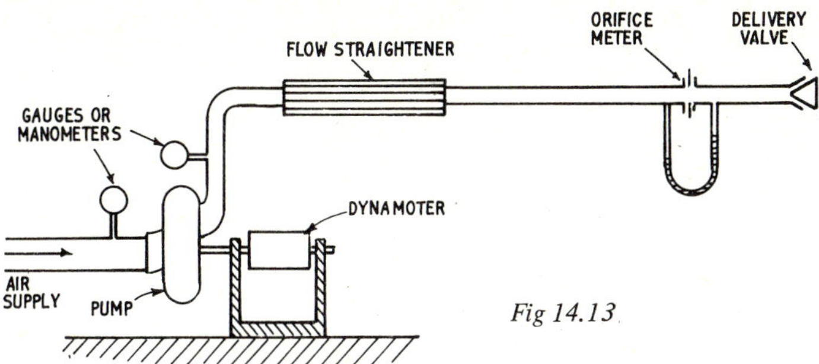

Fig 14.13

to establish the shape of the H-Q curve of the pump, however, although determination of efficiency will usually be quite impossible to any satisfactory level of accuracy.

With compressed air, quite accurate measurement of power input is possible, particularly in a closed circuit system. Likely differences in efficiencies can then be estimated on the basis of the closeness of the respective Reynold's Numbers. $\overline{R}_m$ will normally be lower than $\overline{R}$, when the efficiency of the full size pump can be anticipated as being higher than that of the model. For R_m/R having a value less than 1:15, power measurement is likely to be inaccurate and the difference in efficiencies considerable.

It should be noted that air testing is not confined to reduced scale models and that 1:1 size models (or complete prototype pumps) may be so dealt with. In general, however, this would normally preclude any attempt at all being made to measure power input when using atmospheric air; and if the pump is large the quantity of compressed air required could be prohibitive, making a reduced scale model a more logical choice.

15. Pump Selection Guides

SPECIFIC descriptions of pump characteristics, performance, *etc,* have been covered in previous chapters, together with suitability for specific services, applications or types of fluids, and in particular the size of pump required for specific duty requirements. Since pump types are dealt with separately, this chapter is intended mainly as a guide to the most suitable *type* of pump, or suitable alternatives, under separate requirement headings.

A reservation which applies when listing different pump types as suitable for a particular duty is that other factors may be equally, or even more important, than a specified suitability. Choice should therefore be based on all the factors concerned. This applies particularly where smaller pumps are involved. With large capacity requirements the choice of type generally becomes increasingly limited and tends towards the automatic selection of a suitable type of centrifugal pump. In such cases selection of pump *size* is of primary importance to establish a working point within the acceptable working range or working envelope of the pump in order to achieve optimum efficiency. A further factor, favouring the selection of a centrifugal pump for most duties involving bulk fluid handling, is that the specific speeds realised with direct drives from electric motors correspond to high pump efficiencies.

Standardised Pumps

The description *standardised pump* refers to a design conforming to specified dimensional standards so that different makes of the same pump size will have common dimensions. The main object of this is for savings in time and cost due to standardisation of pipework layouts, together with a reduction in spares requirements. At the same time standardisation may specify an agreed hydraulic grid for pump performance. This results in simplification of design and tender procedures. The widely accepted DIN 24255 standard covers both aspects of standardisation.

Head and Capacity Range

As a starting point, **Fig 15.1** can be consulted to establish likely suitable pump types for working within the head and capacity range required. Limitations which may be imposed by the nature of the fluid to be handled, working conditions or other requirements, can then be investigated in more detail by reference

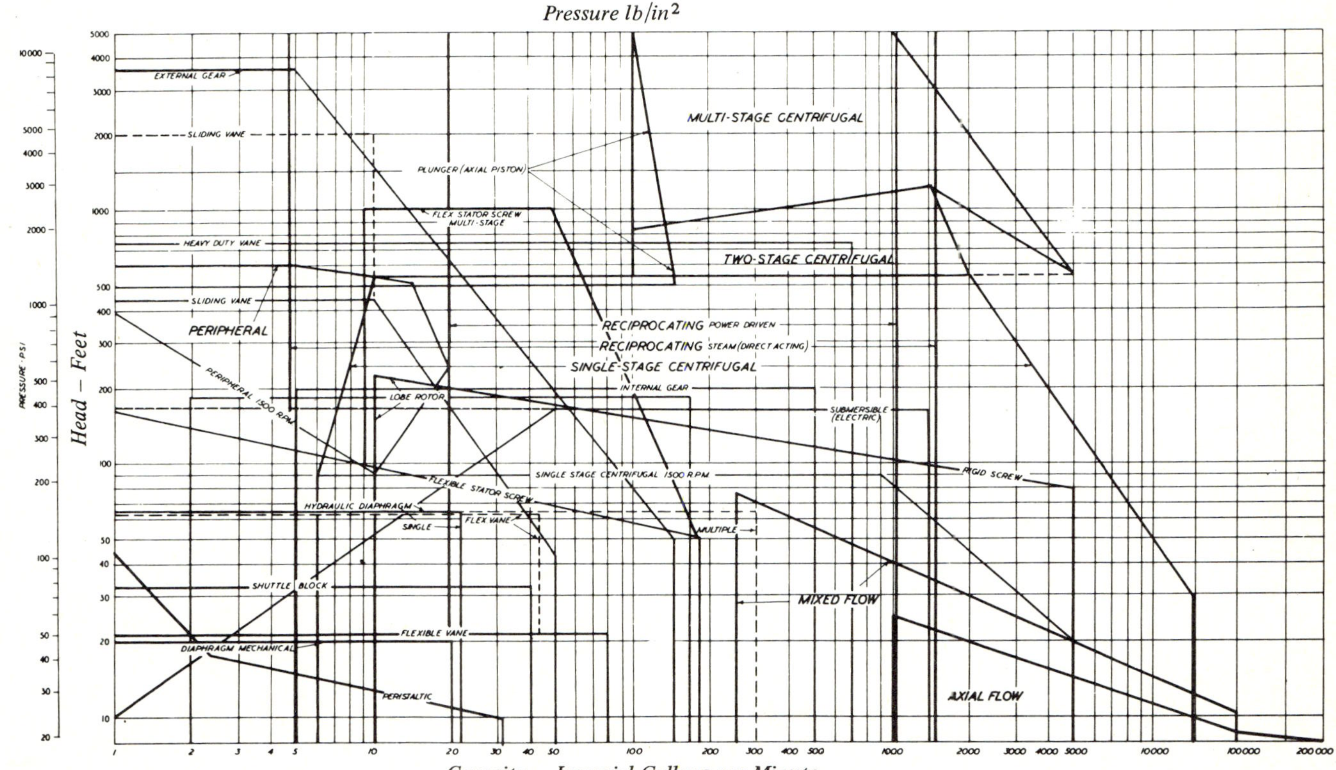

Fig 15.1

Capacity – Imperial Gallons per Minute

It should be noted that the envelopes defining the range of head (or pressure) and capacity available from a particular type does not necessarily mean that a single pump may be capable of developing both maximum capacity and maximum head. Thus a high head (high pressure) pump may be limited to a smaller size; or a high capacity pump may be restricted in pressure range.

to the characteristics of the pump types indicated as possible. This selection chart must be taken as a general guide only, as certain pump types, particularly centrifugal pumps, may be operated successfully well outside their normal working range, although efficiency will be reduced as a consequence.

Capacity Required

Selection on the basis of capacity required can often determine the pump type or types most likely to be suitable for a particular requirement. Capacity ranges, however, overlap with different types of pumps and so this can only be a general guide. Possible alternatives can then be narrowed down by considering other alternatives. (See also **Table I**).

The greatest number of alternatives are usually available in the smaller capacity ranges. Here efficiency and cost may be primary factors. Pump types which are inherently small capacity types will generally show much higher efficiencies than centrifugal pumps where efficiency drops with decreasing size. Cost, on the other hand, may or may not be bound up with efficiency.

Where initial outlay only is significant then a relatively inefficient pump may provide the best answer – *eg* an elementary centrifugal pump which lends itself to mass production by casting or moulding. Where operating costs are the more important, then the pump type would be selected on the basis of highest efficiency at the required capacity – other characteristics being suitable.

Pump sizes are commonly quoted in terms of branch size, but this alone cannot be used as a measure of the capacity of the pump. Thus for the same *branch* size a reciprocating pump may deliver only 15–30% of the discharge of a centrifugal pump; and a typical rotary pump only 10–25% of the discharge of a centrifugal pump. Different *types* of pumps cannot, therefore, be sized on the basis of branch size. Branch size can, however, be used as a very rough guide as to the likely maximum delivery of a conventional design of pump of the same type – see **Table II**.

Head and Capacity Required

Capacity required, together with the total head which the pump must develop, defines the required *working point* of a rotodynamic pump, from which a suitable size of pump can be selected from charts or tables of combined characteristics – see **Chapter 5**. In the case of positive displacement pumps, head is normally a minor factor as regards selection of a suitable pump size, but the head requirements may favour a particular *type* of pump – see **Head or Pressure Required**.

The head-capacity characteristics of a centrifugal pump are dependent on the impeller diameter and rotational speed (or more correctly, specific speed); but also on the specific design of the impeller. Thus, individual pump designs may depart from the general pattern of H-Q characteristics related to specific speed. There is also the possibility that centrifugal pumps with steep H-Q curves may be operated at higher or lower heads than those normally quoted for the working

TABLE I – SELECTION BY CAPACITY

Pump Type	Miniature	Very Small	Small	Moderate	High	Very High	Remarks
Centrifugal Single stage	X*	X*	X	X	X	X	Up to 50 000 gal/min (22 m^3/min) typical
Two-stage			X	X	X		10-5 000 gal/min (50-22 000 l/min) typical
Multi-stage			X	X	X		10-5 000 gal/min (50-22 000 l/min) typical
Self priming			X	X			1-1 500 gal/min (5-7 000 l/min) typical
Regenerative			X	X			Up to 450 gal/min (2 000 l/min) typical
Mixed flow				X	X	X	
Axial flow					X	X	Up to highest attainable
Borehole Shaft driven				X	X		Up to 400 gal/min (2 000 l/min)
Borehole Immersible				X	X	X	Up to 30 000 gal/min (120 000 l/min)
Reciprocating Piston	X		X	X			–
Reciprocating Piston plunger		X	X	X	X		Single and multi-cylinder designs. Small bore large pressure pumps
Direct-acting				X	X		Simplex or Duplex
Radial piston		X	X				Small bore high pressure pumps
Gear	X	X	X				–
Lobe rotor		X	X	X			–
Vane	X	X	X	X			–
Screw (rigid)			X	X	X		Low speeds
Screw, flexible stator			X	X			Inherently low speed type
Diaphragm	X	X	X	X			–
Peristaltic	X	X	X				–
Flexible vane	X	X	X				50 gal/min (225 l/min) usual maximum
Ejectors		X	X	X			Steam or water actuated

*Low efficiency

TABLE II – TYPICAL BRANCH SIZES OF CENTRIFUGAL PUMPS

Suction Inlet in (mm)	4 (100)	6 (150)	12 (300)	24 (600)	36 (900)
Delivery: gal/min lit/min	200–250 900–1100	600–750 2730–3400	3000 14000	10000–12000 45000–55000	25000 115000

TABLE III – SELECTION BY CAPACITY AND HEAD

Type of Pump	Low Capacity			Medium Capacity		High Capacity	
	Low Head	Med Head	High Head	Low Head	High Head	Low Head	High Head
Centrifugal, Single stage	X	X	X	X	X		
Two-stage		X	X	X	X	X	X
Multi-stage			X	X	X	X	X
Self priming			X	X	X		
Regenerative, Single stage		X	X	X			
Multi-stage			X	X	X		
Mixed flow						X	X
Axial flow						X	
Bore hole, Submersible			X		X		X
Bore hole, Immersible			X		X		X
Portable submersible	X		X	X			
Portable immersible	X		X	X			
Reciprocating piston				X	X	X	X
Reciprocating plunger		X	X				
Radial piston					X		
Gear	X	X	X	X	X		
Lobe rotor	X	X		X			
Vane	X	X	X	X			
Screw (rigid)				X	X	X	X
Screw, flexible stator	X	X		X			
Diaphragm	X	X		X			
Peristaltic	X			X			
Flexible vane	X						
Ejectors				X		X	

range (with decrease or increase of capacity, respectively), although this will inevitably result in lowered efficiency.

Table III is a general guide only to suitability of different pump types for different head-capacity requirements, and is the least reliable in the case of centrifugal pumps for the reasons stated above.

Head or Pressure Required (see also Table IV)

Certain pump types are inherently limited as regards the pressure or head they can develop, and thus for high head or high pressure applications are automatically eliminated from selection. In the case of positive displacement pumps a high delivery pressure will increase the input power required and also decrease the actual delivery. This may affect selection of the most suitable pump in two ways, *viz:*

(i) If the increase input power required is not available the driver will be overloaded, or the pump speed will fall, resulting in loss of capacity.

(ii) A slightly larger pump may be required than the size indicated by nominal capacity (capacity at zero delivery pressure) in order that the actual discharge at the actual working pressure corresponds to the capacity required.

It should be noted that the capacity of a particular size of positive displacement pump is variable by adjusting the speed of the pump, but this is not necessarily a practical solution where a constant speed driver is employed.

In the case of rotodynamic pumps, head or pressure cannot be considered as a separate factor except in more general terms – *ie* the higher the head required the lower the specific speed of the pump and where the head required is greater than that which can readily be achieved by a single impeller, two-stage or multistage machines must be employed.

In the case of small submersible pumps (*eg* cellar drainage and sump pumps) powered by constant speed drivers a range of working heads may be specified much wider than the normal working range of a centrifugal pump. Efficiency will vary considerably over the possible working range, the design being a compromise to provide working up to a particular vertical lift. Maximum efficiency will usually be realised at about one half of the specified maximum lift, and is commonly quite low with pumps of this type. This is offset by the convenience of use of such pumps, and the fact that they are not normally called upon to work continuously, and so running costs are not significant.

Suction Lift Required

Maximum suction lift attainable with any pump depends largely on the suction conditions and the NPSH required by the pump employed. Maximum lift likely to be achieved is 27 ft (8.25 m) of water, and practical values attainable are generally lower. Certain pump types are inherently unsuitable for developing

TABLE IV – SELECTION BY HEAD OR PRESSURE

Pump Type	Very Low	Low	Head Mod-erate	High	Very High	Remarks
Centrifugal, Single stage	X	X	X	X		Up to 400-500 ft (120-150 m) max.
Two stage			X	X		Up to highest attainable
Multi stage				X	X	
Self priming	X	X	X	X		Up to 200 ft (60 m)
Regenerative, Single stage	X	X	X			Up to 300 ft (90 m)
Multi stage			X	X		Up to 600 ft (180 m)
Mixed flow	X	X	X			Up to 150 ft (45 m)
Axial flow	X	X				2 to 40 ft (0.6-12 m)
Bore hole submersible			X	X	X	Up to 1200 ft (360 m) depending on number of stages
Bore hole immersible			X	X	X	Up to 1500 ft (450 m) depending on number of stages
Portable sub-mersible	X	X				–
Portable im-mersible	X	X				–
Reciprocating piston		X	X	X		–
Reciprocating, plunger			X	X	X	Up to 20000 lb/in^2 (1400 bar)
Reciprocating, direct acting			X	X		–
Radial piston			X	X	X	Up to 10000 lb/in^2 (700 bar)
Gear		X	X	X		Up to 3500 lb/in^2 (240 bar)
Lobe rotor	X	X	X			Up to 200 ft (60 m)
Vane	X	X	X			Max. 450 up to 2000 lb/in^2 (30–140 bar)
Screw (rigid)			X	X		Up to 600 lb/in^2 (40 bar)
Screw (flexi-ble stator)	X	X	X			Depends on product handled
Diaphragm	X	X				Up to 30-70 lb/in^2 (2-5 bar)
Peristaltic	X	X				Up to 20 lb/in^2 (1-5 bar)
Flexible vane	X	X				Up to 150 lb/in^2 (10 bar) but usually lower
Ejectors		X				–

any practical suction lift, or only very low values. **Table V** gives a general guide as to the suction lift capabilities of different pump types, but this should also be considered in conjunction with self-priming characteristics. Thus a pump which is capable of developing a reasonable suction lift, but is not self-priming,

TABLE V – SELECTION BY SUCTION CHARACTERISTICS

Pump Type	Maximum Suction Lift		Self-Priming	Remarks
	ft	m		
Centrifugal, Miniature	–	–	No	
Centrifugal, Single-stage	27	8.25	No	15 ft (4.5 m) max. typical decreasing with increasing specific speed
Centrifugal, Two-stage	27	8.25	No	” ” ”
Centrifugal, multi-stage	27	8.25	No	” ” ”
Centrifugal, Self-priming	27	8.25	Yes	” ” ”
Regenerative	27	8.25	Yes	” ” ”
Mixed-flow	15	4.5	No	” ” ”
Axial-flow	–	–	No	Usually very low
Borehole submersible	–	–	Yes	Immersed
Borehole immersible	–	–	Yes	Immersed
Immersible/Submersible	–	–	Yes	Immersed
Reciprocating, piston	27	8.25	Yes	22 ft (7 m) max. typical
Reciprocating, plunger	27	8.25	Yes	” ”
Reciprocating, direct-acting	27	8.25	Yes	” ”
Radial piston	27	8.25	Yes	
Radial gear	5 typical	typical	Yes, if wet	Usually very limited – flooded suction usual
Radial vane	6 typical	1.8 typical	Yes (at high speeds)	Self-priming at high speeds
Radial lobe rotor	–	–	No	Depends on rotor form
Radial screw	–	–	–	
Radial diaphragm	27	8.25	Yes	
Radial peristaltic	25	7.5	Yes	
Radial flexible vane	15	4.5	Yes, if wet	
Radial liquid ring	25	7.5	Yes	

may have to be mounted with flooded suction in order to ensure that it starts working. This is particularly so in the case of very small centrifugal pumps (commonly made as submersible units). Larger non-self-priming pumps may be primed manually, or by self-priming devices.

Self-Priming Characteristics

Table V also summarises the self-priming characteristics of the chief types of pumps, together with any specific requirements. Thus, flexible vane pumps have self-priming ability over a reasonable suction lift if the vanes are wetted, *ie* if the casing does not drain completely. Sliding vane pumps, on the other hand, will normally be self-priming in a dry state, but only at high speeds.

Self-priming centrifugal pumps depend on a certain amount of liquid being retained in the casing to initiate self-priming via the auxiliary device fitted. If reduced to a dry state (*eg* when supplied as new, or reassembled after a strip down), the casing will require refilling with liquid initially before the pump will operate as a self-priming unit.

Bearings

Bearing design is the responsibility of the pump manufacturer but the type and disposition of the bearings may affect the application potential and installation requirements. A primary distinction is whether the bearings are mounted outside the casing (external bearings) or inside (internal bearings). Inside bearings will be at least partially, and more usually fully, exposed to contact with the fluid passing through the pump. Thus the choice of an external bearing may be preferred on pumps intended to handle aggressive fluids and pump series may be designed accordingly – *ie* available in similar sizes with the option of external or internal bearings.

Types of bearings may be classified as (i) plain bearings; (ii) ball bearings; (iii) roller bearings and (iv) tilting (pad) or pivoted (segmental).

Plain bearings are widely used on all types and sizes of pumps, being favourable on cost and simplicity of fitting, suitability for high speeds, wide compatability with fluids when wetted and generally quieter operation than rolling bearings. On small pumps simple bushes may be employed, pressed into the pump casing, either of a suitable bearing metal or coated with a low friction surface, or plastic. In the latter case it should be noted that nylon, whilst an excellent bearing material, is subject to swell under continual immersion in aqueous liquids (and also has a higher thermal expansion than metals) and may as a consequence tend to run as a 'tight' bearing with high wear on the shaft; or show high leakage under certain conditions, due to a generous running clearance being adopted. Nylons may be stabilised to reduce this effect.

Larger plain bearings are normally of sleeve type, the sleeve material being a bearing metal or a non-metal. Metal sleeves have the highest load carrying capacity and are generally suitable for wetted bearings where the fluid being pumped has adequate lubricating properties. Water is not excluded, as water lubricated metal bearings are capable of satisfactory performance with suitable choice of materials, provided the water is clean and neutral in characteristics. More usually, however, bearing sleeves for water lubrication are made of rubber, synthetic fibre, laminated thermoset plastic, or plastic.

Rubber is a common choice for larger bearings, with relatively generous clearances and flushing channels in the rubber section, a rubber sleeve being bonded to a metal sleeve (usually bronze) to form the complete bearing sleeve. Rubber bearings are capable of handling dirty as well as clean waters, although other types of water-lubricated bearing sleeves are usually suitable for clean water only. If the wetting fluid is dirty, the bearing sleeves would then normally be lubricated by clean water from a separate supply.

Water lubricated rubber bearings are not normally suitable for use with carbon steel shafts because of the chances of corrosion roughening the shaft. Solutions which may be adopted include the use of a stainless steel shaft (or monel for handling brines); stainless steel or bronze sleeving of the shaft over the bearing area; and hard chromium plating of the shaft, or similar surface treatment. Such shaft protection is not normally necessary with other types of water-lubricated bearing sleeves.

Non-metallic bearings can offer major advantages over metallic bearings for many applications requiring resistance to corrosion or chemical attack, also abrasion; and also with water and other fluids which are non-lurbicants. Materials used include nylon, PTFE, laminated phenolics, acetyl copolymers and ultra-high molecular weight polythene. Even better performance can be achieved in particularly dirty or abrasive conditions with polymer alloys containing a predominant elastomer content. Bearings of this type of hard elastomeric material have the capability of operating dry for a limited period but are normally lubricated by the fluid being pumped (unless incompatible with the material). They can be expected to show superior wear characteristics over bronze, phenolic laminates and carbon-graphite. Many, however, soften at a relatively low temperature (*eg* above 140°F (60°C)). Check both their maximum service temperature and compatibility.

Metallic bearing sleeves are particularly suitable for wetting and lubrication by oil fluids – and are preferable in such cases to non-metallic bearing materials which may be softened or swollen by immersion in oils. Where the fluid is being handled the bearing can be lubricated by a separate oil supply (*eg* oil-ring or oil pump force feed lubrication), or by grease.

Rolling bearings, in general, are to be preferred for most larger pumps, and centrifugal pumps in particular, because of their low friction, compactness and long life. Ball bearings are used for smaller loads and higher speeds, and tapered and spherical roller bearings for heavy loads and lower rotational speeds. The actual type of rolling bearing depends on whether the load is radial, axial (thrust) or combined radial and axial. Plain bearings of the tilting pad type (*eg* Michell or Kingsbury) are also widely used on large machines where there is considerable axial thrust to be absorbed.

Axial thrust loads can be reduced to a minimum on centrifugal pumps (if not eliminated entirely) by incorporating hydraulic balance in the design layout.

This is not possible with mixed-flow or axial flow pumps, although in the former case some appreciable thrust relief can be achieved with balancing holes in the impeller. Bearing design thus becomes increasingly important with such pump types, as the specific speed is increased in order to deal satisfactorily with the thrust loads involved.

With all types of bearings – and with plain bearings in particular – the operating temperature of the bearing will have a considerable influence on bearing performance and life. If the bearing temperature is likely to become excessive – *eg* due to a hot liquid being handled or bearing loads being high – provision may have to be made to cool the bearing. This may be done by fitting a jacket around the bearing through which coolant is circulated; or if the bearing is separately lubricated by water or oil feed, by controlling the temperature or flow rate of the bearing lubricant. The provision of bearing cooling is quite distinct from gland cooling, which is much more commonplace.

Contaminated Fluids (see also Table VI)

Types of pumps for handling fluids containing solids, semi-solids and other contaminants have been discussed in previous chapters. Many such pumps are, in fact, designed for optimum performance handling a particular type of product and would, therefore, normally be a standard choice for such a service. A more general indication of the suitability of different pump types for such services is given in **Table VI**. Remember also that bearings may have to be considered in a pump handling fluids containing abrasives.

Corrosive Fluids

If the product being handled is corrosive this will mainly influence the choice of materials for the wetted parts of the pump, but can also influence the type of pump selected and the choice of pump seals and bearing materials. If the product is also hazardous, zero-leakage or contained-leakage seals may be called for; or a glandless pump.

Packed Gland OR Mechanical Seal?

Many pump manufacturers now offer a packed gland or mechanical seal option on standard pumps. As a general guide where a zero leakage seal is required and/or the product to be contained is aggressive, toxic or extremely volatile, a mechanical seal will normally be a preferred choice, provided the fluid is clean. The more exacting the service the more complex the design of the mechanical seal may need to be, when it can be entirely reliant on the efficiency of ancillary filler, quench or barrier fluid circuits. The principle limitation of a mechanical seal is that it gives no pre-warning of gradual failure; and failure, when it does occur, is total and requires replacement of the whole seal.

TABLE VI – HANDLING CONTAMINATED FLUIDS

Pump Type	Low Concentrations of:			High Concentrations of:		
	Soft Solids	Stringy Solids	Hard Abrasive Solids	Soft Solids	Stringy Solids	Hard Abrasive Solids
Centrifugal:						
Closed impeller	L	X	X	X	X	X
Open impeller	S	S	L	L	L	X
Bladeless	S	S	S	E	S	X
Slurry pump	–	–	S	S	S	E
Regenerative	X	X	X	X	X	X
Mixed flow	S	S	X	X	X	X
Axial flow	S	S	X	X	X	X
Reciprocating:						
Piston	L	X	L	X	X	X
Plunger	X	X	X	X	X	X
Rotary piston	X	X	X	X	X	X
Gear, External	X	X	X	X	X	X
Internal	X	X	X	X	X	X
Lobe rotor	L to X	X	L to X	X	X	X
Sliding vane	X	X	X	X	X	X
Swinging vane	S	S	S	L	L	S
Heavy duty vane	S	L	S	E	L	E
Flexible vane	X	X	L	X	X	X
Diaphragm	E	S	E	E	SE	SE
Tubular diaphragm	S	S	S	S	S	S
Peristaltic	E	E	E	E	E	E
Pneumatic ejector	E	E	L	E	E	L
Single screw (mono)	E	S	E	E	S	E
Fixed screw	L	X	L	X	X	X

Note: For bulk handling the centrifugal pump with suitable impeller is most widely used for stock, stuff, slurries, *etc.* Thus in the case of slurries within the capacity range 100–5 000 gal/min, centrifugal slurry pumps should be used in about 90–95% of all cases, with the remainder shared by diaphragm, single-screw (Mono) and heavy-duty vane pumps.

E = Excellent S = Suitable L = Limited suitability X =Unsuitable

Where some (controlled) leakage is acceptable the packed gland is a robust, readily maintained seal which does give pre-warning of eventual failure. It is also considerably cheaper than a mechanical seal of the same size. This difference becomes greater the larger the seal size required and for very large shafts (*eg* above 5 in (125 mm) a packed gland may be selected on this basis alone.

Pump Efficiencies (see also **Table VIIa** and **VIIb**)

In general the initial efficiency of positive displacement pumps tends to be higher than that of smaller centrifugal pumps; but with increasing pump size

TABLE VIIa – TYPICAL PUMP EFFICIENCIES

Type		Pump Only	Overall
Centrifugal:	large	up to 85–90%	–
	medium	70–75%	–
	small	50–65%	–
	very small	less than 40%	–
	solids handling	45–70%	–
	gravel	50–55%	–
Submersible (centrifugal):			
very small, under 10 gal/min		–	10–25%
small, up to 100 gal/min		–	40%
medium, 100–200 gal/min		–	45–50%
large, 250–400 gal/min		–	50–55%
Radial flow		75–95%	–
Axial flow		75–90%	–
Reciprocating		Up to 90–95%	–
Diaphragm (mechanical)		–	5–20%
External gear		Up to 90%*	–
Internal gear		Up to 80%*	–
Lobe rotor		Up to 80%	–
Shuttle block		10–25%	–
Archimedean screw		up to 75%	–
Single screw (elastomeric stator)		35–65%	–
Triple screw†	low pressures	25–50%	–
	high pressures	55–75%	

†Handling oil fluids 300–3000 Redwood No 1 *

the reverse is normally true. The effect of wear on the two types is quite different. Positive displacement pumps are much more sensitive to increasing clearances due to wear. Doubling of clearances through wear will double the internal leakage and reduce the efficiency accordingly. Doubling of clearances through wear on a centrifugal pump, on the other hand, will be unlikely to reduce efficiency by more than about 10%. It does not follow, however, that handling the same fluid wear rates will be the same on a positive displacement and centrifugal pump.

Efficiency as a parameter must be considered alongside the duty required (*eg* capacity and head), likely varieties in demand, likely pump life and maintenance costs, *etc.* Obviously the greater the volume of product handled the higher the pump efficiency the better as a direct means of minimising energy costs. On the other hand with small capacities involved, other parameters may be significantly more important (*eg* initial cost and pump life).

TABLE VIIb – SMALL SUBMERSIBLE PUMP EFFICIENCIES

Total Head† ft	12 V Battery Operated (70 W*)			240 V Mains Motor (210 W*)		
		Hydraulic Efficiency			Hydraulic Efficiency	
	gal/h	hp	%	gal/h	hp	%
5	250	0.000625	6.7	600	0.0015	5.3
10	225	0.001125	12	550	0.00275	9.7
15	200	0.0015	16	520	0.0039	13.9
20	90	0.0009	9.6	400	0.004	14.2
25	35	0.00088	9.5	350	0.0044	15.6
30				200	0.003	10.6
40				100	0.002	7.1

†Including typical pipeline friction for vertical lift specified
*Maximum

Note: Overall efficiency (%) is calculated on the basis that the power input is the maximum value in each case. This does not necessarily follow. Thus in the case of the battery-powered d.c. motor the motor will speed up and draw less current on lighter loads (*eg* low heads) and slow up and draw increasing current with higher heads.

The main points are:

(i) Small immersible pumps powered by electric motors are generally least efficient when operating against very low heads.

(ii) Maximum efficiency is usually realised at a moderate head (approx. one half the maximum head rating) but will still be low (15% is a good figure for maximum efficiency).

(i) and (ii) are, of course, dependent on the impeller design.

Initial Costs

Initial costs of different types of pumps can vary widely – **Fig 15.2** is a general guide in this respect, comparing the cost of centrifugal and screw pumps over a wide range of similar outputs.

As far as centrifugal pumps are concerned, for standard designs the price for a given capacity and head is usually found to be inversely proportional to the 2/3 power of the speed, *ie*

$$P_2 = P_1 \left(\frac{N_1}{N_2}\right)^{2/3}$$

where:

P_1 = price of pump 1, design speed N_1 rev/min

P_2 = price of pump 2, design speed N_2 rev/min

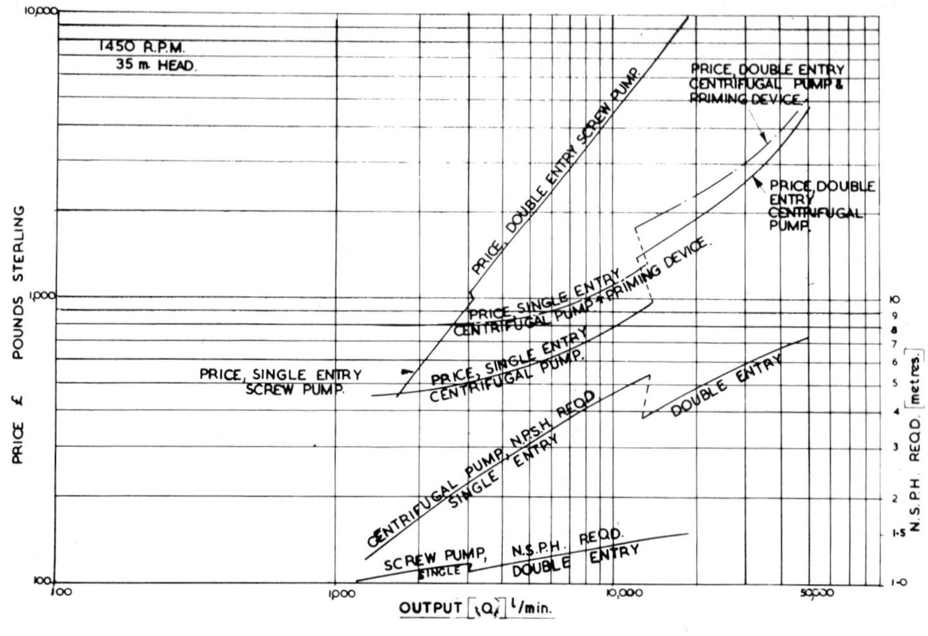

Fig 15.2

Basically, the higher the speed of the pump the lower the initial cost for a required capacity, both for the pump and driver, although this can only be applied directly to a particular class of pump. Thus first consideration should be given to running the selected pump at the highest speed recommended by the manufacturers (bearing in mind any speed limitation imposed by liquid viscosity).

In the case of rotodynamic pumps, consider specific speed as a factor governing the bulk and cost of the pump. Thus for higher speed operation a mixed flow or axial flow pump may prove more satisfactory than a centrifugal pump.

The initial cost of alternative drivers may also be considered in terms of speed and power input required. In general, however, electric motors will usually be most economic as well as covering all power ranges (see also **Table VIII**).

Initial cost of the pump can also be greatly affected by material selection. The true worth of more expensive materials can only be assessed in terms of life achieved and resulting savings in maintenance costs and cost of shutdowns if a cheaper, shorter-life material is used. Simple calculations based on comparative cost/life figures do not give a true picture.

In the case of very small pumps employed in non-critical applications, considerable saving in initial cost may be affected by accepting a limited pump life (enabling cheaper materials and a lower quality pump to be used). The complete pump is then replaced (as a chargeable item) when servicing is required, this often being more economic to both parties than servicing the pump itself.

Running costs are directly proportional to the input power required and the type of driver – see **Table VIII**. The actual cost of pumping, however, is equal to the running cost divided by the efficiency of the pump as worked; and will thus always be higher than the apparent cost. The difference can be extremely significant where a large quantity of fluid is involved, or the pumping plant is run continuously. Thus the lowest pumping costs are achieved with a combination of (i) a driver giving the lowest running costs and (ii) a pump giving the highest possible efficiency for the head and capacity required.

In certain applications – usually associated with low capacities and heads and intermittent operation – pumping costs may be (i) unnecessary to consider; or (ii) negligible. Thus if the service is more important than the cost, or is absorbed or hidden by other running costs, a low efficiency pump may be quite suitable with a saving in initial cost. A typical example is the use of a small (and thus low efficiency) centrifugal pump in domestic washing machines. Similar considerations apply where cost is negligible, a typical example being a low voltage battery-powered electric motor pump where the battery is continuously charged by the generator or alternator of a prime mover working another service.

Maintenance costs add to true running and overall pumping costs and need to be taken into account in all normal pump installations. The true costs involved are:

(i) Cost of actual maintenance period in man hours.

(ii) Cost of replacement parts fitted and handling of returnable parts.

(iii) Cost of loss of pumping time (if applicable).

(iv) Frequency of maintenance required.

Note that whilst the aggregate of (i), (ii) and (iii) represents the total maintenance costs over a period, item (iv) is a 'reciprocal of merit' which is useful for comparative purposes. A low value for (iv) is particularly desirable where continuous operation is important as shut-down intervals for routine inspection and maintenance can be planned for most suitable (least costly) periods.

Overall costs, or the *true* pumping costs, must take into account initial cost, depreciation and any capital interest repayable; direct pumping costs and maintenance costs. The aim in selecting a particular pump for a particular job is to effect the optimum compromise solution as factors favouring low initial cost generally tend to increase all the other costs.

TABLE VIII – PUMP DRIVERS

Driver	Power Range	Power/ Weight	Favourable Factors	Limitations	Fuel Req. per hp-hr	Running cost hp-hr
Manual, hand power	1/10 hp max continuous	–	No cost	Suitable for small hand pumps only	–	–
Animal	1 hp per 1 ton (1000 kg) animal weight	–	Local power source in underdeveloped areas	Only really suitable for water raising pumps	–	–
Windmill	Usually low but possibly up to 7–10 hp in large sizes	–	No fuel required	Depends on wind to work	–	–
Battery powered electric motor	Fractional hp	Fair–good	Independent of fuel or mains supply	Suitable for very small pumps only	–	–
Mains electric motor	All sizes from fractional up to largest required	Fair–good	Lowest initial cost, low to moderate running costs	Dependent on electric power supply	0.9 kW	0.9 x cost per kW hr
Petrol	Fractional hp to 100 hp	Excellent	Compact, flexible, light-	High cost fuel	0.45–0.70	$\frac{\text{cost per gall}}{7.7}$
Paraffin	5–60 hp	Very good	Portable pump sets	Lower cost fuel	0.6–0.9	$\frac{\text{cost per gall}}{9}$
Gas (spark ignition)	4–50 hp	Good	Low cost fuel	May be adapted to use available gas	8–14	10–12 x cost per cu ft
Diesel	About 10 hp upwards	Fair	Portable and fixed pump sets. High reliability, low maintenance	High initial cost	0.3–0.5	$\frac{\text{cost per lb}}{2.2}$

Gas turbine (small)	Up to 50 hp	Exceptional	High power portable and semi-portable pump sets	Very high initial cost	approx 2 lb	$\frac{\text{cost 1 lb fuel}}{0.55}$
Gas turbine (large)	50 hp up	Exceptional	High speed drives	Very high initial cost	approx 1 lb	$\frac{\text{cost 1 lb fuel}}{1.1}$
Steam turbine	Up to 100 hp (single stage)	Excellent	High speed drives for centrifugal pumps	Only economic where steam pressure available	–	–
Air motor (small)	Fractional hp	Excellent	Safe, can be used in hazardous areas	Suitable for very small pumps only	3–4000 cu ft air	3–4 x compressed air cost*
Air motor (large)	Up to 30 hp	Excellent	Safe, can be used in hazardous areas	Limited power range	1800–2000 cu ft air	1.8 – 2 x compressed air cost*
Hydraulic motor	Up to 100 hp	Fair–Good	Only hydraulic motor need be integral with pump	Needs separate hydraulic pump and driver	Depends on driver used	Depends on driver used

*per 1 000 cu ft compressed air

Requisitioning Data

The following data are required when requisitioning a pump or a number of pumps for a particular system.

(i) Nature of service or application.

(ii) Number of pumps required and whether to be used continuously or intermittently.

(iii) Pump type(s) required or preferred (pump user's choice).

(iv) Suction lift, inlet head or NPSH available; also if self-priming pump is required.

(v) Total head or pressure to be developed on delivery side.

(vi) Nature of liquid or product to be handled, if other than clean, cold water.

(vii) Pump layout required, *eg* fixed or portable, horizontal or vertical.

(viii) Site conditions – space available, indoor or outdoor, altitude (if above 500 ft (150 m)), ambient temperature.

(ix) Foundation requirements, if any (to be specified by pump manufacturer).

(x) Type of driver; also if to be directly coupled or indirect drive (*eg* via gearbox, V-belt).

(xi) Starting equipment required; also any required control system.

(xii) Official tests, inspection, shipping requirements, *etc.*

(xiii) Tender receipt/material despatch date required.

Plus any other significant or relevant information.

PRESSURE DROP WITH LAMINAR FLOW

Example: to find the flow loss at 8.0 gal/min through 2 in bore pipe with a fluid of specific gravity 1.0 and viscosity 640 centistokes.
Connect viscosity and flow rate scales. From point thus determined on reference line, connect to pipe bore. Read pressure drop.
Ans: 1 lb/in^2 per foot run of pipe.

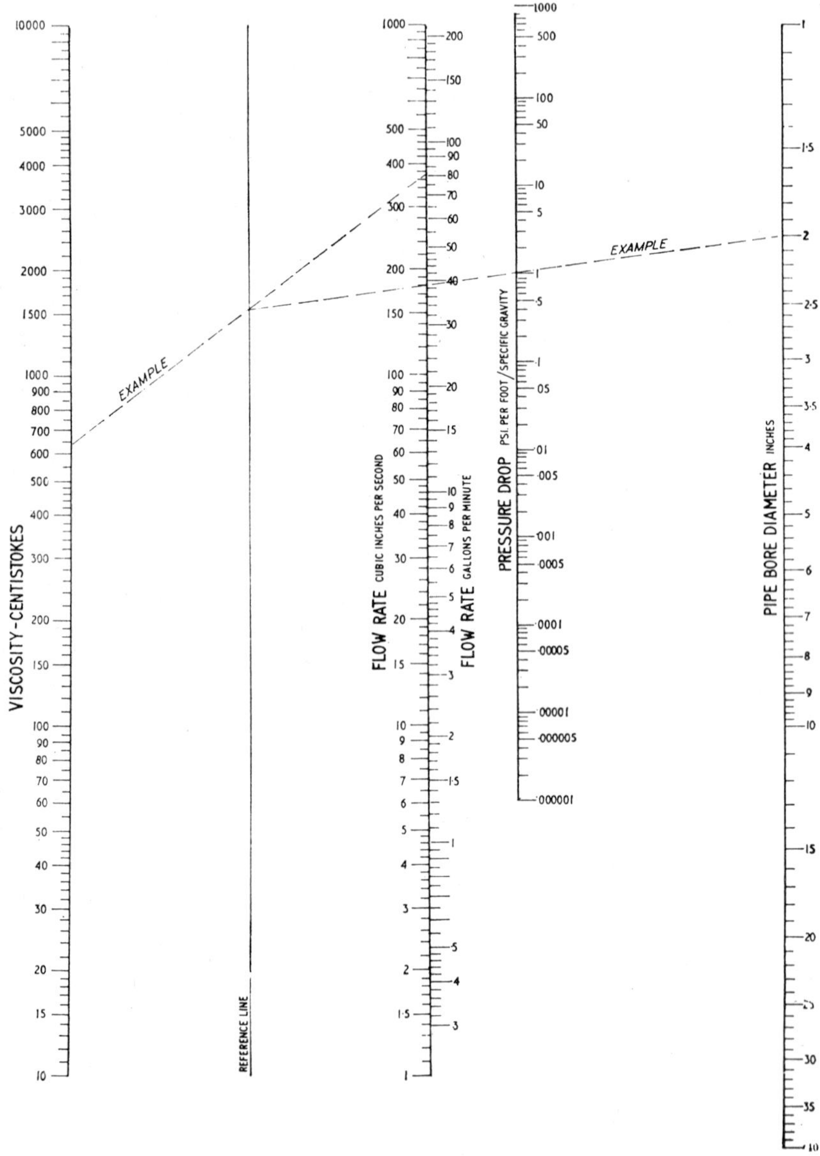

PRESSURE DROP WITH TURBULENT FLOW

Example: given pipe diameter 12 in and mean flow velocity 7.1 ft/sec to find the head loss and pressure drop with a fluid of specific gravity 0.9 and friction factor (previously determined) of 0.031.

Connect friction factor to pipe bore size: then through same point on reference line project from flow velocity value to head loss scale.

Ans: 0.024 ft per foot run.

Project from head loss scale to specific gravity and read pressure drop.

Ans: 0.0095 lb/in^2 per foot run.

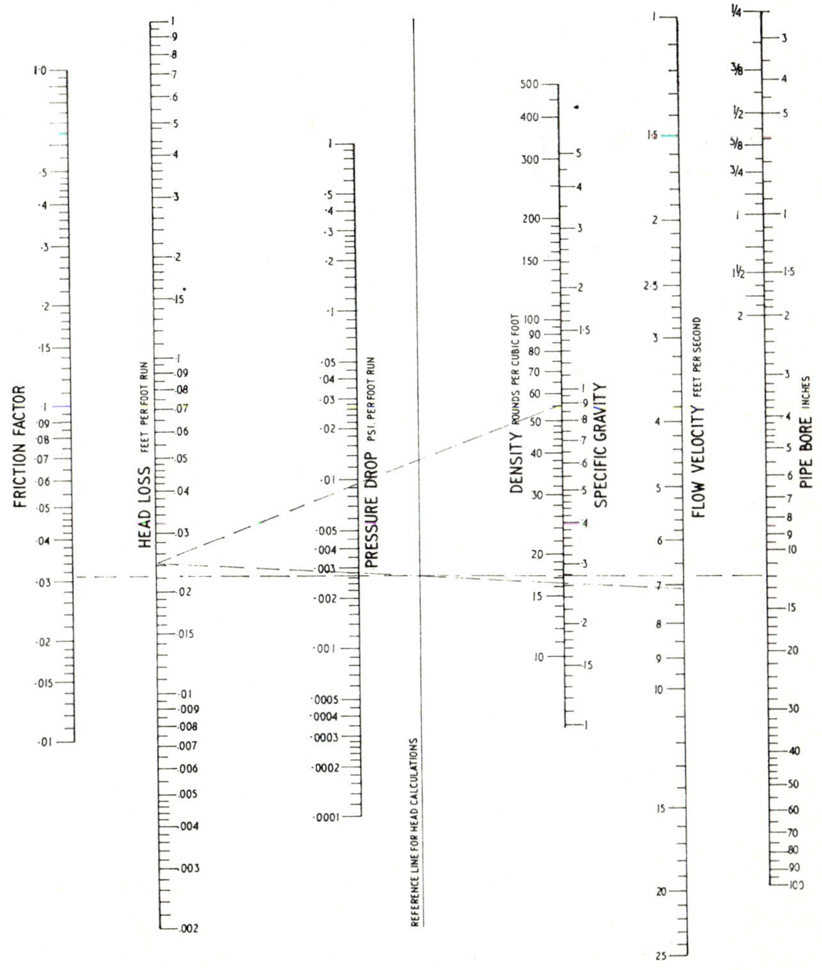

FLOW VELOCITIES COMMONLY USED IN COMMERCIAL PRACTICE

Product	Piping		Flow Velocity ft/sec	m/sec	Remarks
Water	Piston pumps	suction	1.6–5	0.5–1.5	
		delivery	3–6.5	1.0–2.0	
	Feed pumps of steam boilers	suction	1–1.5	0.3–0.5	
		delivery	6.5–8	2.0–2.5	The piping is selected according to its length. A lower velocity is chosen with long piping, a higher velocity for short piping (does not apply to delivery of liquids containing solid particles)
	Piping for condensate and sludge	suction	1–1.5	0.3–0.5	
		delivery	3–6.5	1.0–2.0	
	Gravel, sand and other drifted substances	delivery	1.5–6.5	0.5–2.0	
	Piping for cold water	delivery	3–10	1.0–3.0	
	Piping for cold water	suction			
	up to 2 in dia. (50 mm)	max	3.25	1.0	
	up to 4 in dia. (100 mm)	max	4.25	1.3	
	up to 8 in dia. (200 mm)	max.	5.5	1.7	
	above 8 in dia.(200 mm)		6.5	2.0	
	Piping for cooling water	suction	2.25–5	0.7–1.5	
		delivery	3–6.5	1.0–2.0	
	Pressure water		50–100	15.0–30.0	In special cases up to 16 ft/sec (5 m/sec)
	Delivery piping in mines		3–5	1.0–1.5	
	Supply to water turbines		10	3.0	Low head
			10–23	3.0–7.0	High head
	Municipal water mains, main feed pipings		3–6.5	1.0–2.0	
	Municipal water system		1.5–4	0.5–1.2	Normally 2 to 2.3 ft/sec (0.6 to 0.7 m/sec)
Petrol oils	Benzol, gas oil		3–4	1.0–2.0	According to viscosity
	Heavy		1.5–4	0.5–2.0	
Hydrocarbons		suction	1–2.5	0.3–0.8	
Air	Low-pressure piping		40–50	12–15	
	High-pressure piping		65–80	20–25	
Steam	For steam lines up to 4 MPa		65–130	20–40	Velocities must be chosen economically according to the length of the piping
	High-pressure steam		100–200	30–60	
	Superheated steam		130–260	39–80	
	Low-pressure heating steam		33–50	10–15	
	Exhaust steam		50–130	15–40	

RECOMMENDED PIPE MATERIALS & FLOW VELOCITIES FOR CHEMICAL LIQUIDS

Liquid	Material of Piping	Flow Velocity ft/sec	m/sec
Benzene	steel	6.0	1.8
Bromine	glass	4.0	1.2
Ammonia	steel	6.0	1.8
Dibromomethylene	glass	4.0	1.2
Dichloroethylene	glass	6.0	1.8
Ethylene glycol	steel	6.0	1.8
Sodium hydroxide (Ni 0–30%)	steel (Ni)	6.0	1.8
Sodium hydroxide (30–50%)	steel (Ni)	5.0	1.5
Sodium hydroxide (50–73%)	steel (Ni)	4.0	1.2
Calcium chloride	steel	4.0	1.2
Carbon tetrachloride	steel	4.0	1.2
Chlorine	steel	5.0	1.5
Chloroform	copper, steel	6.0	1.8
Sodium chloride (solution without solid particles)	steel	5.0	1.5
Sodium chloride (solution with solid particles)	Monel metal, Ni	6–15	1.8–4.5
Hydrochloric acid	rubber-coated	5.0	1.5
Sulphuric acid 88–93%	alloy steel	4.0	1.2
Sulphuric acid 93–100%	steel	4.0	1.2
Methyl chloride	steel	6.0	1.8
Perchloroethylene	steel	6.0	1.8
Propylene glycol	steel	5.0	1.5
Styrene	steel	6.0	1.8
Trichloroethylene	steel	6.0	1.8
Sea water	rubber-coated	5–8	1.5–2.5
Vinyl chloride	steel	6.0	1.8
Vinylidene chloride	steel	6.0	1.8

APPROXIMATE VISCOSITY CONVERSIONS & RELATED CENTRIFUGAL PUMP SIZING

Kinematic Viscosity Stokes	Kinematic Viscosity Centistokes	Redwood No. 1 Seconds	Saybolt Universal Seconds	Engler Seconds	Engler Degrees	Redwood Admiralty Seconds	Saybolt Furol Seconds	Barbey Fluidity	Minimum Size Centrifugal Pump in (mm)
0.01	1	29.0	31.0	51.3	1.00	–	–	6200	None
0.02	2	30.9	33.5	57.5	1.12	–	–	3100	
0.03	3	33.0	36.2	62.6	1.22	–	–	2067	
0.04	4	35.3	39.1	67.2	1.31	–	–	1550	
0.05	5	37.9	42.3	71.3	1.39	–	–	1240	
0.06	6	40.5	45.5	75.9	1.48	–	–	1033	
0.07	7	43.2	48.7	80.1	1.56	–	–	886	
0.08	8	46.0	52.0	84.7	1.65	–	–	775	
0.09	9	48.8	55.4	89.3	1.74	–	–	689	
0.1	10	51.7	58.6	93.9	1.83	–	–	620	
0.2	20	85.0	97.5	147	2.87	9	15.0	310	¾–1 (20–25)
0.3	30	123	141	209	4.07	12	18.5	207	1–1¼ (25–32)
0.4	40	163	186	274	5.33	16	22.2	153	1¼–1½ (32–40)
0.5	50	203	231	340	6.61	20	26.0	124	1½–2 (40–50)
0.6	60	244	277	406	7.90	24	30.5	103	1½–2 (40–50)
0.7	70	284	323	473	9.21	28	35.0	88.6	2–2½ (50–65)
0.8	80	324	370	559	10.5	32	39.5	77.5	2–2½ (50–65)
0.9	90	364	416	606	11.8	36	44.0	68.9	2–2½ (50–65)
1	100	405	462	677	13.2	41	48.5	62.0	2–3¼ (50–80)
2	200	810	924	1350	26.3	81	94.7	31.0	3¼–4 (80–100)
3	300	1215	1386	2030	39.5	122	141	20.7	5–6 (125-150)
4	400	1620	1848	2700	52.6	162	188	15.5	6–7 (150-175)
5	500	2025	2310	3580	65.8	203	235	12.4	7–8 (175-200)
6	600	2430	2772	4060	78.9	243	282	10.3	8–10 (200-250)
7	700	2835	3234	4730	92.1	284	329	8.9	8–10 (200-250)
8	800	3240	3696	5390	105	324	376	7.8	10–12 (250-300)
9	900	3645	4158	6060	118	365	423	6.9	10–12 (250-300)
10	1000	4050	4620	6770	132	405	470	6.2	12–14 (300-350)
20	2000	8100	9240	13500	263	810	940	3.10	15–18 (400-450)
30	3000	12150	13860	20300	395	1215	1410	2.07	
40	4000	16200	18480	27000	526	1620	1880	1.55	Positive displacement pump required
50	5000	20250	23100	33800	658	2025	2350	1.24	
60	6000	24300	27720	40600	789	2430	2820	1.03	
70	7000	28350	32340	47300	921	2835	3290		
80	8000	32400	36960	53900	1050	3240	3760		
90	9000	36450	41580	60600	1180	3645	4230		
100	10000	40500	46200	67700	1316	4050	4700		

ABSOLUTE VISCOSITIES IN CENTIPOISE

FOR SPECIFIC GRAVITY OF **0.8**	FOR SPECIFIC GRAVITY OF **0.9**	FOR SPECIFIC GRAVITY OF **1.0**	FOR SPECIFIC GRAVITY OF **1.1**	FOR SPECIFIC GRAVITY OF **1.2**	FOR SPECIFIC GRAVITY OF **1.3**	FOR SPECIFIC GRAVITY OF **1.4**
16,800	18,900	21,000	23,100	25,200	27,300	29,400
15,100	17,000	18,900	20,800	22,680	24,560	26,440
13,440	15,100	16,800	18,500	20,180	21,820	23,500
11,750	13,230	14,700	16,180	17,640	19,100	20,590
10,080	11,340	12,600	13,860	15,120	16,480	17,630
8,400	9,450	10,500	11,550	12,600	13,650	14,700
7,560	8,500	9,450	10,400	11,350	12,300	13,240
6,800	7,650	8,500	9,350	10,200	11,050	11,900
5,880	6,620	7,350	8,090	8,830	9,560	10,300
5,040	5,670	6,300	6,940	7,560	8,200	8,830
4,200	4,720	5,250	5,780	6,300	6,830	7,350
3,400	3,820	4,250	4,680	5,100	5,530	5,950
2,520	2,840	3,150	3,460	3,780	4,090	4,410
1,760	1,980	2,200	2,420	2,640	2,860	3,080
1,560	1,750	1,950	2,150	2,340	2,530	2,730
1,360	1,530	1,700	1,870	2,040	2,210	2,380
1,200	1,350	1,500	1,650	1,800	1,950	2,100
1,040	1,170	1,300	1,430	1,560	1,690	1,820
840	945	1,050	1.150	1,260	1,370	1,470
680	765	850	935	1,020	1,100	1,190
505	567	630	694	756	820	883
336	378	420	462	504	546	588
176	198	220	242	264	286	308
156	175	195	214	234	253	273
136	153	170	187	204	221	238
120	135	150	165	180	195	210
104	117	130	143	156	169	182
84	94	105	109	126	136	147
68	77	85	94	102	111	119
50	57	63	69	76	82	88
34	38	42	46	50	55	59
18	20	22	24	26	29	31
15	17	19	21	23	25	27
14	15	17	19	20	22	24
12	14	15	17	18	20	21
8	9	10	11	12	13	14
6	6	7	8	8	9	10
3	4	4	4	5	5	6

TABLE IA

BASIC CAPACITY CONVERSIONS

CUBIC CENTIMETRES	CUBIC INCHES	CUBIC FEET	IMPERIAL GALLONS	US GALLONS	US BARRELS	LITRES	CUBIC METRES
1	.061024	—	—	—	—	—	—
16.3871	1	—	—	—	—	—	—
—	1728	1	6.232	5.914	.17811	28.3161	.028317
—	277.42	.160544	1	1.20095	.028594	4.54596	.0045461
—	231.0	.1337	.83267	1	.02381	3.78533	.0037854
—	9702	5.6146	34.9726	42	1	158.984	.1590
1000.028	61.024	.035316	.219975	.26397	.00629	1	.00100003
—	—	35.30565	219.969	264.17	6.29	999.972	1

TABLE IB

PRACTICAL DELIVERY CONVERSIONS

cc sec	cu.in/ sec	cu.in/ min	gallons/ min	US gallons/ min	gallons/ hour	US gallons/ hour	Litres/ min	Litres/ hour
1	.061	3.66	.013	.016	.8	.95	.061	3.64
16.4	1	60	.2163	.26	13	15.6	1(approx)	59
.273	.0167	1	.0036	.0043	.216	.258	.0164	1
75.5	4.62	277.4	1	1.2	60	72	4.55	272.8
63	3.85	231	.83	1	.50	.60	3.8	227.3
1.26	.077	4.62	.0167	.02	1	1.2	.076	4.55
1.05	.064	3.85	.014	.017	.83	1	.063	3.8
16.4	1	61	.22	.264	.00367	0044	1	60
.273	.017	1	.0035	.0042	.22	.264	.0167	1

COMPATIBILITY OF RUBBER LININGS

	COMPATIBILITY*	RUBBERS					EBONITES	
		Natural	Butyl	Neoprene	Nitrile	CSM	Natural Rubber or SBR	Nitrile
	Maximum Service Temperature	70–80 °C	90–110 °C	100 °C	100 °C	90 °C	65–70 °C	110 °C
Acids	Acetic	X	L	S	L	S	S	S
	Chromic (50%)	X	X	X	X	L	X	X
	Hydrochloric	S	S	L	L	S	S	S
	Hydrocyanic	S	S	S	S	S	S	S
	Hydrofluoric	S	S	L	L	S	S	L
	Nitric (8%)	X	S	X	X	S	X	X
	Nitric (20%)	X	L	X	X	S	X	X
	Nitric (40%)	X	X	X	X	S	X	X
	Phosphoric (80%)	S	S	S	L	S	S	S
	Sulphuric (20%)	S	S	S	L	S	S	S
	Sulphuric (35%)	S	S	S	L	S	S	S
	Sulphuric (50%)	S	S	L	L	S	S	S
Salts & Bases	Ammonia (25%)	S	S	S	S	S	S	S
	Ammonium hydroxide	S	S	L	S	S	S	S
	Ferric chloride	L	S	L	S	S	S	S
	Silver nitrate	L	S	S	L	S	S	S
	Zinc chloride	S	S	S	S	S	S	S
Miscellaneous Organic	Acetone	S	S	X	X	X	S	S
	Alcohols	S	S	L	S	S	S	S
	Detergents	S	S	S	S	S	S	S
	Ethers	X	L	L	S	S	X	X
	Ethylene glycol	S	S	S	S	S	S	S
	Methyl chloride	X	L	X	X	X	X	X
	Murral oils	X	X	L	S	S	L	S
	Phenol	L	L	X	X	X	S	S
	Soaps	S	S	S	S	S	S	S
	Vegetable oils	X	S	L	S	S	L	S

* S = Satisfactory or suitable
X = Not suitable
L = Limited suitability

VAPOUR PRESSURE OF WATER

Temp °C	Vapour Pressure			Temp °C	Vapour Pressure		
	mm Mercury	lb/in^2	Head (ft)		mm Mercury	lb/in^2	Head (ft)
0	4.579	0.09	0.21	50	92.51	1.82	4.2
1	4.926	0.10	0.23	51	97.20	1.91	4.4
2	5.294	0.10	0.24	52	102.09	2.01	4.6
3	5.655	0.11	0.26	53	107.20	2.11	4.9
4	6.101	0.12	0.28	54	112.51	2.22	5.1
5	6.543	0.13	0.30	55	118.04	2.33	5.4
6	7.013	0.14	0.32	56	123.80	2.44	5.6
7	7.513	0.15	0.35	57	129.82	2.55	5.9
8	8.045	0.16	0.37	58	136.08	2.67	6.2
9	8.609	0.17	0.39	59	142.60	2.80	6.5
10	9.209	0.18	0.41	60	149.38	2.94	6.8
11	9.844	0.19	0.44	61	156.43	3.07	7.1
12	10.518	0.21	0.48	62	163.77	3.21	7.4
13	11.231	0.22	0.51	63	171.38	3.36	7.8
14	11.987	0.24	0.55	64	179.31	3.51	8.1
15	12.788	0.25	0.58	65	187.54	3.68	8.5
16	13.634	0.27	0.62	66	196.09	3.84	8.9
17	14,530	0.29	0.66	67	204.96	4.02	9.3
18	15.477	0.30	0.69	68	214.17	4.20	9.7
19	16.477	0.32	0.74	69	223.73	4.38	10.2
20	17.535	0.34	0.78	70	233.7	4.57	10.7
21	18.650	0.37	0.84	71	243.9	4.78	11.1
22	19.827	0.39	0.90	72	254.6	5.00	11.6
23	21.068	0.41	0.95	73	265.7	5.21	12.1
24	22.377	0.44	1.02	74	277.2	5.43	12.5
25	23.756	0.47	1.08	75	289.1	5.67	13.1
26	25.209	0.49	1.15	76	301.4	6.00	13.9
27	26.739	0.53	1.22	77	314.1	6.15	14.2
28	28.349	0.56	1.3	78	327.3	6.42	14.8
29	30.043	0.59	1.35	79	341.0	6.68	15.5
30	31.824	0.62	1.4	80	355.1	6.92	16.1
31	33.695	0.66	1.5	81	369.7	7.25	16.8
32	35.663	0.70	1.6	82	384.9	7.55	17.5
33	37.729	0.74	1.7	83	400.6	7.85	18.2
34	39.898	0.78	1.8	84	416.8	8.16	18.9
35	42.175	0.83	1.9	85	433.6	8.50	19.6
36	44.563	0.88	2.0	86	450.9	8.83	20.4
37	47.067	0.93	2.1	87	468.7	9.15	21.2
38	49.692	0.98	2.3	88	487.1	9.55	22.1
39	52.442	1.03	2.4	89	506.1	9.92	22.9
40	55.324	1.09	2.5	90	525.8	10.3	23.8
41	58.34	1.15	2.7	91	546.1	10.7	24.8
42	61.50	1.21	2.8	92	567.0	11.1	25.6
43	64.80	1.27	2.9	93	588.6	11.6	26.8
44	68.26	1.34	3.1	94	610.9	12.0	27.7
45	71.88	1.41	3.3	95	633.9	12.4	28.6
46	75.65	1.49	3.5	96	657.6	12.9	29.8
47	79.60	1.57	3.7	97	682.1	13.4	31.0
48	83.71	1.65	3.8	98	707.3	13.9	32.1
49	88.02	1.73	4.0	99	733.2	14.3	33.0
				100	760	14.7	34.0

CAST IRON AND CAST STEEL SPECIFICATIONS

Material	British Standard Specification	Equivalent American Specification	Ult Tensile tons/in²	Stress N/mm²	Min Yield Stress tons/in²	N/mm²	Elongation min %	Red. in area % min	IZOD ft lb min	Approx Brinell Range min	Approx Brinell Range max
A Cast iron	BS1452 Grade 12	—	12	—	—	—	—	—	—	180	210
A Cast iron	BS1452 Grade 14	—	14	—	—	—	—	—	—	190	220
A Cast iron	BS1452 Grade 17	—	17	—	—	—	—	—	—	200	230
B Itestion Grade 3	BS1452 Grade 20	—	20	—	—	—	—	—	—	220	250
C Itestion Grade 4	BS1452 Grade 23	—	23	—	—	—	—	—	—	250	300
D 0.15/0.25%C Cast steel	BS592 Grade A BS1617 Grade B BS1504-161 Grade A	ASTM A 27 Grade 65-30 — ASTM A 216 Grade WCA	28/32	430/490	15	230	22	30	15	115	145
E 0.25/0.35%C Cast steel	BS592 Grade B BS1504-161 Grade B	ASTM A 27 Grade 70-36 ASTM A 216 Grade WCB	32/35	490/540	17	260	18	35	15	135	165
F 0.35/0.4% Cast steel	BS592 Grade C BS1504-101 Grade C	— ASTM A 148 Grade 80-40	35/40	540/620	19	295	14	30	10	145	175
G 0.4/0.45%C Cast steel	BS1760 Grade A	—	40/45	620/690	21	525	12	20	—	165	205
H 0.45/0.5%C	BS1760 Grade A	—	45/50	690/770	24	370	8	15	—	180	220
J 0.18/0.25%C-1.2/1.7MN	BS1456 Grade A	ASTM A 148 Grade 80/50	35Min	540/690	22	320	16	25	25	150	185
K 0.25/0.33%C-1.2/1.7MN	BS1456 Grade B1	—	40/45	620/690	24	370	13	30	20	165	215
L 0.25/0.35% NiCrMo	BS1458 Grade A	—	45/55	690/850	32	495	11	20	25	201	255

APPLICATION

A These grades of iron castings are employed where castings with smooth clean skin, with good machinability, are required. Repetition machine moulded castings from mounted patterns.

B Superior qualities of strength, wear resistance and damping capacity.

C High duty castings, machine cut gears, castings of variable section demanding uniform structure, high wear resistance and strength.

D General engineering and structural castings, easily machined and specially suitable for including in fabricated weldments.

E General engineering castings with increased tensile strength for parts under pressure, valves *etc.*

F Wear resistant castings for gear wheels and general application.

G Wear resistant castings for gear wheels and general applications. Suitable for flame or induction hardening.

H Wear resistant castings giving maximum hardness when locally hardened. Suitable for flame or induction hardening.

J Castings subject to shock loading where high impact resistance is required and applications for low temperature service.

K Similar to above where higher tensile strength with good impact is required and applications for low temperature service.

L Castings requiring high fatigue and wear resistance.

INDEX